High-Speed Decoders for Polar Codes

Pascal Giard • Claude Thibeault
Warren J. Gross

High-Speed Decoders
for Polar Codes

 Springer

Pascal Giard
Institute of Electrical Engineering
École Polytechnique Fédérale de Lausanne
Lausanne, VD, Switzerland

Claude Thibeault
Department of Electrical Engineering
École de Technologie Supérieure
Montréal, QC, Canada

Warren J. Gross
Department of Electrical and Computer
    Engineering
McGill University
Montréal, QC, Canada

ISBN 978-3-319-86699-4   ISBN 978-3-319-59782-9   (eBook)
DOI 10.1007/978-3-319-59782-9

© Springer International Publishing AG 2017
Softcover reprint of the hardcover 1st edition 2017
This work is subject to copyright. All rights are reserved by the Publisher, whether the whole or part of the material is concerned, specifically the rights of translation, reprinting, reuse of illustrations, recitation, broadcasting, reproduction on microfilms or in any other physical way, and transmission or information storage and retrieval, electronic adaptation, computer software, or by similar or dissimilar methodology now known or hereafter developed.
The use of general descriptive names, registered names, trademarks, service marks, etc. in this publication does not imply, even in the absence of a specific statement, that such names are exempt from the relevant protective laws and regulations and therefore free for general use.
The publisher, the authors and the editors are safe to assume that the advice and information in this book are believed to be true and accurate at the date of publication. Neither the publisher nor the authors or the editors give a warranty, express or implied, with respect to the material contained herein or for any errors or omissions that may have been made. The publisher remains neutral with regard to jurisdictional claims in published maps and institutional affiliations.

Printed on acid-free paper

This Springer imprint is published by Springer Nature
The registered company is Springer International Publishing AG
The registered company address is: Gewerbestrasse 11, 6330 Cham, Switzerland

I wanna go fast!
Ricky Bobby

Preface

Origin

The majority of this book was initially published as a Ph.D. thesis, a thesis nominated for the *Prix d'excellence de l'Association des Doyens des Études Supérieures au Québec (ADESAQ)* by the Electrical and Computer Engineering department of McGill University.

Scope

Over the last decades we have gradually seen digital circuits take over applications that were traditionally bastions of analog circuits. One of the reasons behind this tendency is our ability to detect and correct errors in digital circuits—circuits making computations with discrete signals as opposed to continuous ones. This ability led to faster and more reliable communication and storage systems. In some cases it enabled things that we thought might have never been possible, e.g., reliable communication with a probe that is located many light years away from our planet.

Right after the Second World War, Claude Shannon created a new field—information theory—in which he defined the limit of reliable communications or storage. In his seminal work, Shannon defined what he calls the channel capacity [60], the bound that many researchers have tried to achieve or even approach ever since. Shannon's work does not tell us how this limit can be reached.

While Reed-Solomon (RS) and Bose-Chaudhuri-Hocquenghem (BCH) codes have good error-correction performance and are in widespread use even today, it's not until the discovery of turbo codes [12] in the 1990s that error-correcting codes approaching the channel capacity were found. Indeed, while Low-Density Parity-Check (LDPC) codes—initially discovered in the 1960s by

Robert Gallager [16]—can also be capacity approaching, their decoding algorithm was too complex for the time and thus were not used until they were independently rediscovered by David McKay in 1997 [39].

The discovery of turbo and LDPC codes greatly rejuvenated the field of error correction. Often used in conjunction with a RS or a BCH code, standards that feature a turbo or a LDPC code are omnipresent. Nowadays, each home contains at least tens of decoders for these codes. They are used in a plethora of applications such as video broadcasting, wireless and wired communications (e.g., WIFI and Ethernet), and data storage.

The latest findings on the road to achieving channel capacity are polar codes. Invented by Arıkan in 2008 [6] and further refined in 2009 [7], this new class of error-correcting codes, contrary to LDPC and turbo codes, has an explicit—nonrandom—construction making the implementation of their encoders and decoders simpler than that of LDPC or turbo codes. Polar codes exploit the channel polarization phenomenon by which the probability of correctly estimating codeword bits tends to either 1 (completely reliable) or 0.5 (completely unreliable). These probabilities get closer to their limit as the code length increases when a recursive construction is used. Under the low-complexity Successive-Cancellation (SC) decoding algorithm, polar codes were shown to achieve the symmetric capacity of memoryless channels as their length tends to infinity. The complexity of the SC algorithm is low but its sequential nature translates in high-latency and low-throughput decoder implementations. To overcome this, new decoding algorithms derived from SC were introduced, most notably [4] and [55]. These algorithms exploit the recursive construction of polar codes along with the a priori knowledge of the code structure. Fast Simplified Successive Cancellation (Fast-SSC), the algorithm described in [55], integrates the Simplified Successive Cancellation (SSC) algorithm described in [4]; thus this book builds upon the former.

Fast-SSC represented a significant improvement over the previous algorithms and led to the first hardware decoder for polar codes achieving a throughput greater than 1 Gbps. However, the optimization presented therein targeted high-rate codes. As low-rate codes are omnipresent in modern wireless communications, it was evident that it would be beneficial to have a closer look at potential improvements for such codes.

In Software-Defined Radio (SDR) applications, researchers and engineers have yet to fully harness the error-correction capability of modern codes. Many are still using classical codes [13, 63] as implementing low-latency high-throughput—exceeding 10 Mbps of information throughput—software decoders for turbo or LDPC codes is very challenging. The irregular data access patterns featured in turbo and LDPC decoders make efficient use of Single-Instruction Multiple-Data (SIMD) extensions present on today's processors difficult. To overcome the difficulty of efficiently accessing memory while decoding one frame and still achieve a good throughput, software decoders resorting to inter-frame parallelism (decoding multiple independent frames at the same time) are often proposed [30, 66, 69]. Inter-frame parallelism comes at the cost of higher latency, as many frames have to be buffered before decoding can be started. Even with a split layer approach to LDPC decoding where intra-frame parallelism can be applied, the latency remains high at multiple milliseconds on a recent desktop processor [23]. On the other hand, polar codes are well suited for software implementation as their decoding algorithms feature regular memory access patterns.

While the future 5G standards are still in the works, many documents mention the requirement of peak per-user throughput greater than 10 Gbps. Regardless of the algorithm, the state of polar decoder implementations when our research started offered much lower throughput. The fastest SC-based decoder had a throughput of 1.2 Gbps at a clock frequency of 106 MHz [55]. The fastest decoder implementation based on the Belief Propagation (BP) decoding algorithm—an algorithm with higher parallelism than SC—had an average 4.7 Gbps throughput when early termination was used with a clock frequency of 300 MHz [49]. It was evident that a minor improvement over the existing architectures was unlikely to be sufficient to meet the expected throughput requirements of future wireless communication standards.

The book presents a comprehensive evaluation of decoder implementations of polar codes in hardware and in software. In particular, the work exposes new trade-offs in latency, throughput, and complexity, in software implementations for high-performance computing and General-Purpose Graphical Processing Units (GPGPUs), and hardware implementations using custom processing elements, full-custom Application-Specific Integrated Circuits (ASICs), and Field-Programmable Gate Arrays (FPGAs).

The book maintains a tutorial nature clearly articulating the problems that polar decoder implementations are facing, and incrementally develops various novel solutions. Various design approaches and evaluation methodologies are presented and defended. The work advances the state of the art while presenting a good overview of the research area and future directions.

Organization

This book consists of six chapters. Chapter 1 reviews polar codes, their construction, representations, and encoding and decoding algorithms. It also briefly goes over results for the state-of-the-art decoder implementations from the literature.

In Chap. 2, improvements to the state-of-the-art low-complexity decoding algorithm are presented. A code construction alteration method with human-guided criteria is also proposed. Both aim at reducing the latency and increasing the throughput of decoding low-rate polar codes. The effect on various low-rate moderate-length codes and implementation results are discussed.

Algorithm optimization at various levels leading to low-latency high-throughput decoding of polar codes on modern processors is introduced in Chap. 3. Bottom-up optimization and efficient use of SIMD instructions available on both embedded-platform and desktop processors are proposed in order to parallelize the decoding of a frame, reduce latency, and increase throughput. Strategies for efficient implementation of polar decoders on GPGPU are also presented. Implementation results for all three types of modern processors are discussed.

A family of hardware architectures utilizing unrolling is presented in Chap. 4 showing that polar decoders can achieve extremely high-throughput values and retain moderate complexity. Implementations for various rates and code lengths are presented for FPGA and ASIC. The results are compared with the state of the art.

Expending from the previous chapter, Chap. 5 introduces a method to enable the use of multiple code lengths and rates in a fully unrolled polar decoder architecture. This novel method leads to a length- and rate-flexible decoder while retaining the very high speed typical to those decoders. ASIC results are presented for two versions of a multi-mode decoder and compared against the state-of-the-art decoders.

Lastly, conclusions about this book are drawn in Chap. 6 and a list of suggested future research topics is presented.

Audience

This book is aimed at error-correction researchers who heard about polar codes—a new class of provably capacity achieving error-correction codes—and who would like to learn about practical decoder implementation challenges and trade-offs in either software or hardware. As polar codes just got accepted to protect the control channel in the next-generation mobile communication standard (5G) developed by the 3GPP [40], this includes engineers who will have to implement decoders for such codes. Some prior experience in software or hardware implementation of high performance signal processing systems is an asset but not mandatory. The book can also be used by SDR practitioners looking into implementing efficient decoders for polar codes, or even hardware engineers designing the backbone of communication networks. Additionally, it can serve as reading material in graduate courses notably covering modern error correction.

Lausanne, VD, Switzerland   Pascal Giard
Montreal, QC, Canada   Claude Thibeault
Montreal, QC, Canada   Warren J. Gross

Acknowledgements

Many thanks to my friend and former colleague Gabi Sarkis. A lot of this work would have been tremendously more difficult to nearly impossible without his help. His algorithmic, software and hardware skills, his vast knowledge, and his insightful comments were all of incredible help. Furthermore, his willingness to cooperate led to very fruitful collaborations stirring both of us up and helping me to remain motivated during the harder times.

I would also like to thank Alexandre J. Raymond, Alexios Balatsoukas-Stimming, and Carlo Condo who helped me in one way or another. Thanks to Samuel Gagné, Marwan Kanaan, and François Leduc-Primeau for the interesting discussions we had during our downtime.

I am grateful for the financial support I got from the Fonds Québécois de la Recherche sur la Nature et les Technologies, the fondation Pierre Arbour, and the Regroupement Stratégique en Microsystèmes du Québec.

Finally, I would like to thank my beautiful boys Freddo and Gouri as well as my wonderful and beloved Joëlle. Their patience, support, and indefectible love made this possible. Countless times, Joëlle had to sacrifice or take everything on her shoulders so that I could pursue my dreams. I am very grateful and privileged that she stayed by my side.

Lausanne, Vaud, Switzerland   Pascal Giard

Contents

Acronyms

ASIC	Application-Specific Integrated Circuit
AVX	Advanced Vector eXtensions
AWGN	Additive White Gaussian Noise
BCH	Bose-Chaudhuri-Hocquenghem
BEC	Binary Erasure Channel
BER	Bit-Error Rate
BP	Belief Propagation
BPSK	Binary Phase-Shift Keying
BSC	Binary Symmetric Channel
CC	Clock Cycle
CPU	Central Processing Unit
CRC	Cyclic Redundancy Check
DRAM	Dynamic Random-Access Memory
Fast-SSC	Fast Simplified Successive Cancellation
FEC	Forward Error Correction
FER	Frame-Error Rate
FPGA	Field-Programmable Gate Array
GPGPU	General Purpose GPU
GPU	Graphical Processing Unit
I/O	Input/Output
IoT	Internet of Things
LDPC	Low-Density Parity Check
LHS	Left Hand Side
LLR	Log-Likelihood Ratio
LTE	Long-Term Evolution
LUT	Look-Up Table
ML	Maximum Likelihood
ML-SSC	Simplified Successive Cancellation with Maximum-Likelihood nodes
OFDM	Orthogonal Frequency-Division Multiplexing
PE	Processing Element
RAM	Random-Access Memory

RHS	Right Hand Side
RS	Reed-Solomon
RTL	Register-Transfer Level
SC	Successive Cancellation
SDR	Software-Defined Radio
SIMD	Single Instruction Multiple Data
SIMT	Single Instruction Multiple Threads
SoC	System on Chip
SPC	Single Parity Check
SP-SC	Semi-Parallel Successive Cancellation
SRAM	Static Random-Access Memory
SSC	Simplified Successive Cancellation
SSE	Streaming SIMD Extensions
SSSE	Supplemental Streaming SIMD Extensions
TP-SC	Two-Phase Successive Cancellation

Chapter 1
Polar Codes

Abstract This chapter reviews polar codes, their construction, representations, and encoding and decoding algorithms. It also briefly goes over results for the state-of-the-art decoder implementations from the literature.

1.1 Construction

Polar codes exploit the channel polarization phenomenon to achieve the symmetric capacity of a memoryless channel as the code length increases ($N \rightarrow \infty$). A polarizing construction where $N = 2$ is shown in Fig. 1.1a. The probability of correctly estimating bit u_1 increases compared to when the bits are transmitted without any transformation over the channel W. Meanwhile, the probability of correctly estimating bit u_0 decreases. The polarizing transformation can be combined recursively to create longer codes, as shown in Fig. 1.1b for $N = 4$. As the $N \rightarrow \infty$, the probability of successfully estimating each bit approaches either 1 (perfectly reliable) or 0.5 (completely unreliable), and the proportion of reliable bits approaches the symmetric capacity of W [7].

To construct an (N, k) polar code, the $N - k$ least reliable bits, called the frozen bits, are set to zero and the remaining k bits are used to carry information. Figure 1.2a illustrates non-systematic encoding of an $(8, 4)$ polar code, where the frozen bits are indicated in gray and $a_0, \ldots, a_3$ are the $k = 4$ information bits. Encoding is carried out by propagating $\boldsymbol{u} = u_0^7$ from left to right, through the graph of Fig. 1.2a.

The locations of the information and frozen bits are based on the type and conditions of W. Unless specified otherwise, in this book we use polar codes constructed according to [61]. The generator matrix, G_N, for a polar code of length N can be specified recursively so that $G_N = F_N = F_2^{\otimes \log_2 N}$, where $F_2 = \left[\begin{smallmatrix} 1 & 0 \\ 1 & 1 \end{smallmatrix}\right]$ and $\otimes$ is the Kronecker power. For example, for $N = 4$, G_N is

$$G_4 = F_2^{\otimes 2} = \left[\begin{array}{c|c} F_2 & 0 \\ \hline F_2 & F_2 \end{array}\right] = \begin{bmatrix} 1 & 0 & 0 & 0 \\ 1 & 1 & 0 & 0 \\ 1 & 0 & 1 & 0 \\ 1 & 1 & 1 & 1 \end{bmatrix}.$$

© Springer International Publishing AG 2017
P. Giard et al., *High-Speed Decoders for Polar Codes*,
DOI 10.1007/978-3-319-59782-9_1

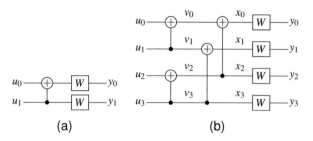

Fig. 1.1 Construction of polar codes of lengths (**a**) $N = 2$ and (**b**) $N = 4$

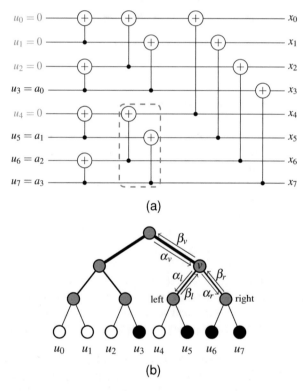

Fig. 1.2 Non-systematic $(8, 4)$ polar code represented as a (**a**) graph and as a (**b**) decoder tree

In matrix form, non-systematic encoding can be represented as $x = uG_N$, where u is a N-bit row vector containing the bits to be encoded in the information bit locations. When polar codes were initially proposed, bit-reversed indexing was used. While this changes the bit ordering for both encoding and decoding, the error-correction performance remains unaffected. This change translates into multiplying the generator matrix by the bit-reversal permutation matrix B_N [7] (or $\tilde{\Pi}_N$ [6]), so that $G_N = B_N F_N$. In this book, natural indexing is used unless stated otherwise.

1.2 Tree Representation

A polar code of length N is the concatenation of two constituent polar codes of
length $N/2$ [7]. Therefore, binary trees are a natural representation of polar codes
[4]. Figure 1.2 illustrates the tree representation of an $(8, 4)$ polar code. In Fig. 1.2a,
the frozen bits are labeled in gray while the information bits are in black. The
corresponding tree, shown in Fig. 1.2b, uses white and black leaf nodes to denote
these bits, respectively. The gray nodes of Fig. 1.2b correspond to concatenation
operations shown in Fig. 1.2a. Moving up in the decoder tree corresponds to the
concatenation of constituent codes. For example, the concatenation operation circled
in blue in Fig. 1.2a corresponds to the node labeled v in Fig. 1.2b.

1.3 Systematic Coding

Encoding schemes for polar codes can be either non-systematic, as shown in
Figs. 1.1b and 1.2a, or systematic as discussed in [8]. Systematic polar codes
offer better Bit-Error Rate (BER) than their non-systematic counterparts; while
maintaining the same Frame-Error Rate (FER). Furthermore, they allow the use
of low-complexity rate-adaptation techniques such as code shortening method
proposed in [38]. Flexible low-complexity systematic encoding of polar codes is
discussed at length in [53, 58].

Figure 1.3 shows an example of the low-complexity systematic encoding scheme
proposed in [53, 58]. It comprises two non-systematic encoding passes and a
bit masking operation in between. For a $(8, 4)$ polar code, a N-bit vector $\boldsymbol{u} =
[0, 0, 0, a_0, 0, a_1, a_2, a_3]$, where $a_0, \ldots, a_3$ are the $k = 4$ information bits, enters the
first non-systematic encoder from the left. Then, using bit masking, the locations
corresponding to frozen bits are reset to 0 before propagating the updated vector

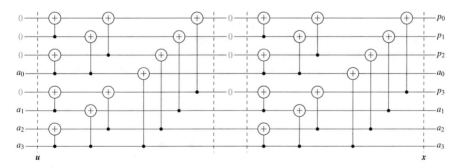

Fig. 1.3 Low-complexity systematic encoding of a $(8, 4)$ polar code

through the second non-systematic encoder. The end result is a N-bit vector $x = [p_0, p_1, p_2, a_0, p_3, a_1, a_2, a_3]$, where $p_0, \ldots, p_3$ are the $N - k = 4$ parity bits and $a_0, \ldots, a_3$ are the k information bits.

This encoding scheme was proven to be correct under certain conditions, conditions that are always met when a construction method leading to polar codes with a good error-correction performance is used e.g. [61]. In this book, systematic polar codes are used.

1.4   Successive-Cancellation Decoding

In SC decoding, the decoder tree is traversed depth first, selecting left edges before backtracking to right ones, until the size-1 frozen and information leaf nodes. The messages passed to child nodes are Log-Likelihood Ratios (LLRs); while those passed to parents are bit estimates. These messages are denoted α and β, respectively. Messages to a left child l are calculated by the f operation using the min-sum algorithm:

$$\begin{aligned} \alpha_l[i] &= f(\alpha_v[i], \alpha_v[i + N_v/2]) \\ &= \text{sign}(\alpha_v[i])\text{sign}(\alpha_v[i + N_v/2]) \min(|\alpha_v[i]|, |\alpha_v[i + N_v/2]|), \end{aligned} \tag{1.1}$$

where N_v is the size of the corresponding constituent code and α_v the LLR input to the node.

Messages to a right child are calculated using the g operation

$$\begin{aligned} \alpha_r[i] &= g(\alpha_v[i], \alpha_v[i + N_v/2], \beta_l[i]) \\ &= \begin{cases} \alpha_v[i + N_v/2] + \alpha_v[i], & \text{when } \beta_l[i] = 0; \\ \alpha_v[i + N_v/2] - \alpha_v[i], & \text{otherwise,} \end{cases} \end{aligned} \tag{1.2}$$

where β_l is the bit estimate from the left child.

Bit estimates at the leaf nodes are set to zero for frozen bits and are calculated by performing threshold detection for information ones. After a node has the bit estimates from both its children, they are combined to generate the node's estimate that is passed to its parent

$$\beta_v[i] = \begin{cases} \beta_l[i] \oplus \beta_r[i], & \text{when } i < N_v/2; \\ \beta_r[i - N_v/2], & \text{otherwise,} \end{cases} \tag{1.3}$$

where $\oplus$ is modulo-2 addition (XOR).

1.5 Simplified Successive-Cancellation Decoding

As mentioned above, a polar code is the concatenation of smaller constituent codes. Instead of using the SC algorithm on all constituent codes, the location of the frozen bits can be taken into account to use more efficient, lower complexity, algorithms on some of these constituent codes. In [4], decoder-tree nodes are split into three categories: Rate-0, Rate-1, and Rate-R nodes.

1.5.1 Rate-0 Nodes

Rate-0 nodes are subtrees whose leaf nodes all correspond to frozen bits. We do not need to use the SC algorithm to decode such a subtree as the exact decision, by definition, is always the all-zero vector.

1.5.2 Rate-1 Nodes

These are subtrees where all leaf nodes carry information bits, none are frozen. The maximum-likelihood (ML) decoding rule for these nodes is to take a hard decision on the input LLRs:

$$\beta_v[i] = \begin{cases} 0, & \text{when } \alpha_v[i] \geq 0; \\ 1, & \text{otherwise.} \end{cases} \tag{1.4}$$

With a fixed-point representation, this operation amounts to copying the most significant bit of the input LLRs.

1.5.3 Rate-R Nodes

Lastly, Rate-R nodes, where $0 < R < 1$, are subtrees such that leaf nodes are a mix of information and frozen bits. These nodes are decoded using the conventional SC algorithm until a Rate-0 or Rate-1 node is encountered.

As a result of this categorization, the SSC algorithm trims the SC decoder tree for a $(8, 5)$ polar code shown in Fig. 1.4a into the one illustrated in Fig. 1.4b. Rate-1 and Rate-0 nodes are shown in black and white, respectively. Gray nodes represent Rate-R nodes. Trimming the decoder tree leads to a lower decoding latency and an increased decoder throughput.

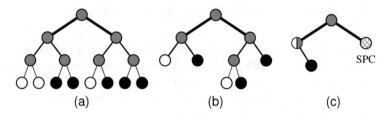

SPC

(a)   (b)   (c)

Fig. 1.4 Decoder trees corresponding to the (**a**) SC, (**b**) SSC and (**c**) Fast-SSC decoding algorithms

1.6   Fast-SSC Decoding

The Fast-SSC decoding algorithm extends both SC and SSC and further prunes the decoder tree by applying low-complexity decoding rules when encountering certain types of constituent codes.

Three functions—F, G and *Combine*—are inherited from the original SC algorithm. They correspond to (1.1), (1.2) and (1.3), respectively. Fast-SSC also integrates the decoding algorithms for the Rate-1 and Rate-0 nodes of the SSC algorithm.

However, for some Rate-R nodes corresponding to constituent codes with specific frozen-bit locations, a decoding algorithms with lower latency than SC decoding is used. These special cases are:

1.6.1   Repetition Codes

Repetition codes are constituent codes where only the last bit is an information bit. These codes are efficiently decoded by calculating the sum of the input LLRs and using threshold detection to determine the result that is then replicated to form the estimated bits:

$$\beta_v[i] = \begin{cases} 0, & \text{when } \left(\sum_{i=0}^{N_v-1} \alpha_v[i]\right) \geq 0; \\ 1, & \text{otherwise,} \end{cases}$$

where N_v is the number of leaf nodes.

1.6.2   SPC Codes

Single Parity Check (SPC) codes are constituent codes where only the first bit is frozen. The corresponding node is indicated by the cross-hatched orange pattern in Fig. 1.4c. The first step in decoding these codes is to calculate the hard decision of each LLR

$$\beta_v[i] = \begin{cases} 0, & \text{when } \alpha_v[i] \geq 0; \\ 1, & \text{otherwise}, \end{cases} \tag{1.5}$$

and then calculating the parity of these decisions

$$\text{parity} = \bigoplus_{i=0}^{N_v-1} \beta_v[i]. \tag{1.6}$$

If the parity constraint is unsatisfied, the estimate of the bit with the smallest LLR magnitude is flipped:

$$\beta_v[i] = \beta_v[i] \oplus \text{parity}, \text{ where } i = \arg\min_j(|\alpha_v[j]|). \tag{1.7}$$

1.6.3 Repetition-SPC Codes

Repetition-SPC codes, or RepSPC codes, are codes whose left constituent code is a repetition code and the right an SPC one. They can be speculatively decoded in hardware by simultaneously decoding the repetition code and two instances of the SPC code: one assuming the output of the repetition code is all 0's and the other all 1's. The correct result is selected once the output of the repetition code is available. This speculative decoding also provides speed gains in software.

1.6.4 Other Operations

The Fast-SSC algorithm introduces other types of operations with the aim of reducing the number of memory accesses, and thus of reducing the latency.

Notably the *G0R* and *C0R* (or *Combine_0R*) operations are special cases of the *G* and *Combine* operations, respectively (1.2) and (1.3), where the left child is a frozen node i.e. β_l is known a priori to be the all-zero vector of length N_v.

Figure 1.4c shows the tree corresponding to a Fast-SSC decoder.

1.7 Other SC-Based Decoding Algorithms

Other SC-based algorithms were published where multiple bits are estimated at a time. The next two sections present a brief overview of the most notable ones.

1.7.1 ML-SSC Decoding

ML-SSC [57] expands on SSC by using an exhaustive-search ML decoder to decode rate-R codes once their length and dimension fall below a resource-constrained threshold. The general rule for ML decoding with LLR inputs is given by

$$\beta_v = \arg \max_{x \in \mathscr{C}} \sum_i (1 - 2x_i)\alpha_{v_i}; \tag{1.8}$$

where α_v is the LLR input and $\mathscr{C}$ is the list of codewords of the constituent code.

1.7.2 Hybrid ML-SC Decoding

The hybrid ML-SC decoding algorithm [37] partitions the polar code graph into M partitions, where each is decoded using an SC decoder until stage $\log_2 M$ is reached. At that point different rules are used based on the location and count of frozen bits. Instead of conducting an exhaustive search, the ML decoder is simplified by taking advantage of the special structure of polar codes. Nonetheless, no approximations are made and these rules are thus equivalent to the ML decoding rule (1.8).

In the hybrid ML-SC algorithm, SC decoders first produce M LLR values that are used by the following ML decoder section to estimate M bits. These estimated bits are then used to calculate the next M LLR values according to (1.2), and so on. Since the progression of the decoding process and the operations applied in hybrid ML-SC are the same as those of ML-SSC, the former can be seen as a special case of the latter.

1.8 Other Decoding Algorithms

Besides SC-based algorithms, other algorithms can be used to decode polar codes. On one hand, there are prohibitively complex algorithms, like sphere [28] or linear-programming [22] decoding, practically restricted to short polar codes because of their complexity with regard to code length. On the other hand, there are algorithms that may turn out to be interesting but that did not get much attention yet, in particular the BP and the List-based algorithms. The former is interesting because of its intrinsic high level of parallelism and the latter has great potential because it can significantly improve the error-correction performance of short- to moderate-length polar codes.

1.8.1   Belief-Propagation Decoding

The BP algorithm is a well-known algorithm that has been very successfully applied to decode LDPC codes. It was shown in [26] that it can be adapted to decode polar codes as well. BP decoding of a polar code can be seen as applying a flooding decoding schedule to the graph representation of a polar code as opposed to a serial schedule such as the one used in SC-based decoding.

LLRs are iteratively propagated in the graph until a stopping criterion is met. This criterion can either be an early-stopping criterion [73] or simply a fixed maximum number of iterations. Threshold detection is then applied to the resulting LLRs to generate the codeword estimate.

It was shown that BP decoding may require a very large number of iterations to achieve the same error-correction performance as SC. Figure 1.5 shows an example where BP decoding of a $(2048, 1723)$ polar code requires at least 100 iterations of a flooding schedule to match the performance of SC decoding. At equal error-correction performance, even a fully-parallel BP decoder has a greater latency than an SC decoder.

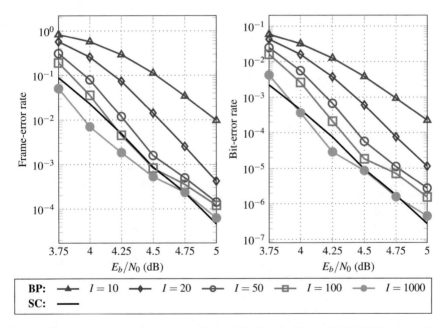

Fig. 1.5 Error-correction performance of BP and SC decoding for a $(2048, 1723)$ polar code, where I is the maximum number of iterations. Data from [53] and used with author's permission

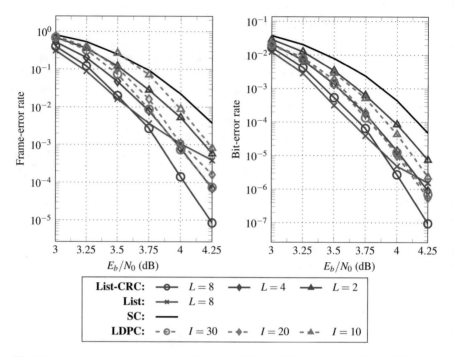

Fig. 1.6 Error-correction performance of List, List-CRC and SC decoding of a (2048, 1723) polar code versus that of the (1944, 1620) 802.11n LDPC code. L is the maximum number of candidate codewords and I is the maximum number of iterations

1.8.2 List-Based Decoding

In list-based decoding algorithms, several decoding paths are explored using an SC-based algorithm and a constrained list of the L-best candidate codewords is built. These L-best candidates are determined by calculating reliability metric for each of the explored paths. It was shown in [62] that list decoding a polar code concatenated with a Cyclic Redundancy Check (CRC)—List-CRC decoding—greatly improves the error-correction performance over list decoding of a polar code alone. This improvement is significant enough to have polar codes exceed the performance of LDPC codes of similar length and rate.

Figure 1.6 shows the error-correction performance of List-based decoding of a (2048, 1753) polar code. The performance of SC decoding as well as that of the (1944, 1620) LDPC code from the 802.11n WIFI standard are included for comparison. A maximum of 10, 20 or 30 iterations of offset min-sum BP decoding with a flooding schedule were used for the LDPC code. All List-CRC decoding curves are for a 16-bit CRC.

In a list-based decoder, the L paths can either be processed in parallel using up to L SC-based decoders or serially by time-multiplexing the use of $M < L$ SC-based decoders. The former results in increased hardware complexity, and the latter in higher latency and lower throughput decoders. Efficient hardware implementations of list-based decoders for polar codes capable of achieving a throughput greater than 5 Gbps was an open problem when we started this book and so it remains to this day.

1.9 SC-Based Decoder Hardware Implementations

Since this book proposes SC-based hardware decoders, this section briefly reviews the notable and state-of-the-art SC-based hardware decoders from the literature. But first, a module central to most SC-based decoders is briefly discussed: the Processing Element (PE).

1.9.1 Processing Element for SC Decoding

In SC decoding, the soft-value messages are calculated using (1.1) and (1.2). Very early on, a block integrating both calculations was proposed and designated as a PE [35]. Later, it was proposed to use a special PE for the last calculation stage in order to estimate 2 bits simultaneously [42, 51, 72]. As will be shown below, a different approach was taken in the Fast-SSC implementation.

1.9.2 Semi-Parallel Decoder

The Semi-Parallel Successive-Cancellation (SP-SC) decoder, first proposed in [33] to be later improved in [51], puts a constraint on the number of implemented PEs. It was observed that, in the line SC decoder [34]—which represented the state of the art at the time—the $N/2$ PEs were only used twice per decoded frame.

It was shown in [33] that by implementing $P = 64$ PEs instead of $N/2$, a SP-SC decoder would reach 97% of the speed of a line decoder when decoding a polar code of length $N = 2^{11}$, while reducing the hardware complexity by an order of magnitude.

1.9.3 Two-Phase Decoder

The Two-Phase Successive-Cancellation (TP-SC) decoder [48] mainly aimed at reducing the memory requirements in a SC-based decoder. As its name suggests, the decoding process is broken down into two parts. Phase 1 and phase 2 are for constituent codes of lengths $N_v > \sqrt{N}$ and $N_v \leq \sqrt{N}$, respectively.

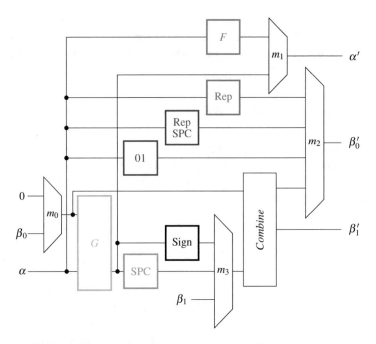

Fig. 1.7 Architecture of the data processing unit proposed in [55]

A pipelined-tree architecture is used to increase the clock frequency and memory is saved by not saving all intermediate LLRs to Random-Access Memory (RAM), instead they are recalculated when then are needed again.

1.9.4 Processor-Like Decoder or the Original Fast-SSC Decoder

The Fast-SSC implementation of [55] proposed a decoder architecture that contained, at its core, what resembles a processor. As illustrated in Fig. 1.7, the processor featured all the modules required to implement the nodes and operations described in Sects. 1.4, 1.5 and 1.6.

The decoder could be configured at run time to load a set of instructions corresponding to given polar code. At the time however, the code length was fixed at synthesis time. It was later improved to support any polar code of length $N \leq N_{\text{max}}$ [58].

Table 1.1 Post-fitting results for SC-based decoder implementations

Implementation	N	LUTs	Regs.	RAM (kbits)	f (MHz)
SP-SC [33]	1024	2888	1388	11.9	196
SP-SC [50]		2618	1292	13.8	169
TP-SC [48]		1940	748	7.1	239
SP-SC [33]	16 384	29 897	17 063	184.1	113
SP-SC [50]		2769	1230	206.6	168
TP-SC [48]		7815	3006	114.6	230
Fast-SSC [55]		25 219	6529	285.3	106
SP-SC [33]	32 768	58 480	33 451	364.3	66
SP-SC [50, 51]		3263	1304	411.6	167
Fast-SSC [55]		25 866	7209	536.1	108

Table 1.2 Latency and information throughput for SC-based decoder implementations

Implementation	Code (N, k)	Latency (CCs)	Latency (μs)	Info. T/P (Mbps)
SP-SC [33]	(1024, 512)	2304	12	44
SP-SC [50]		2604	15	33
TP-SC [48]		2656	11	56
SP-SC [33]	(16 384, 14 746)	34 304	304	48
SP-SC [50]		43 772	261	57
TP-SC [48]		41 600	181	106
Fast-SSC [55]		1433	14	1091
SP-SC [33]	(32 768, 29 492)	69 120	1047	28
SP-SC [50, 51]		88 572	530	56
Fast-SSC [55]		2847	26	1081

1.9.5 Implementation Results

In this section, results for implementations of the decoders discussed above are presented. All results are for the Altera Stratix IV EP4SGX530-KH40C2 FPGA. Table 1.1 shows the resource usage and execution frequency while Table 1.2 presents latency and throughput results for a few polar codes of various lengths and rates. It should be noted that the TP-SC decoder implementation of [48] lacks the buffers required to sustain its throughput.

Chapter 2
Fast Low-Complexity Hardware Decoders for Low-Rate Polar Codes

Abstract In this chapter, we show how the state-of-the-art low-complexity decoding algorithm can be improved to better accommodate low-rate codes. More constituent codes are recognized in the updated algorithm and dedicated hardware is added to efficiently decode these new constituent codes. We also alter the polar code construction to further decrease the latency and increase the throughput with little to no noticeable effect on error-correction performance. Rate-flexible decoders for polar codes of length 1024 and 2048 are implemented on FPGA and ASIC. Over the previous FPGA work, they are shown to have from 22 to 28% lower latency and 26 to 34% greater throughput when decoding low-rate codes. On 65 nm ASIC CMOS technology, the proposed decoder for a $(1024, 512)$ polar code is shown to compare favorably against the state-of-the-art ASIC decoders. With a clock frequency of 400 MHz and a supply voltage of 0.8 V, it has a latency of 0.41 μs and an area efficiency of 1.8 Gbps/mm^2 for an energy efficiency of 77 pJ/info. bit. At 600 MHz with a supply of 1 V, the latency is reduced to 0.27 μs and the area efficiency increased to 2.7 Gbps/mm^2 at 115 pJ/info. bit.

2.1 Introduction

While the Fast-SSC [55] algorithm represents a significant improvement over the previous decoding algorithms, the work in [55] and the optimization presented therein targeted high-rate codes. In this chapter, we propose modifications of the Fast-SSC algorithm and a code construction alteration process targeting low-rate codes. We present results using the proposed methods, algorithms and implementation. These results show a 22–28% latency reduction and a 22–28% throughput improvement with little to negligible coding loss for low-rate moderate-length polar codes.

The rest of this chapter is organized as follows. Section 2.2 discusses polar code construction alteration along with our proposed method leading to improved latency and throughput of a hardware decoder. In Sect. 2.3, modifications to the original Fast-SSC algorithms are proposed in order to further reduce the latency and increase the decoding throughput. Sections 2.4 and 2.5 present the implementation details along with the detailed results on FPGA. Section 2.5 also provides ASIC results for our proposed decoder decoding a $(1024, 512)$ polar code for a comparison against state-of-the-art ASIC decoders from the literature. Finally, Sect. 2.6 concludes this chapter.

2.2 Altering the Code Construction

2.2.1 Original Construction

As mentioned in Chap. 1, a good polar code is constructed by selecting which bits to freeze, according to the type of channel and its conditions [7, 44, 61, 65]. Figure 2.1 shows the decoder tree corresponding to the $(1024, 512)$ polar code constructed using the technique of [61] where only the node types defined in Table 2.1 are used with the same constraints of [55]. The polar code was optimized for an E_b/N_0 of 2.5 dB. The *F*, *G*, *G_0R*, *Combine* and *Combine_0R* blocks are constrained to a maximum of $P = 512$ inputs meaning that, for nodes with a length $N_v > P$, $\lceil N_v/P \rceil$ cycles are required. The *Rep*, *RepSPC* and 01 blocks are all executed in one clock cycle. Finally, the SPC-based nodes—0SPC and RSPC—use pipelining and require $\lceil N_v/P \rceil + 4$ clock cycles. Thus the decoding latency to decode the tree of Fig. 2.1 using the algorithm and implementation of [55] is 220 Clock Cycles (CCs) and the information throughput is 2.33 bits/CC.

Altering a polar code to further trim the decoder tree can result in a significant latency reduction, without affecting the code rate. By making these modifications however, the error-correction performance is degraded. Although, as will be shown in the next section, the impact can be small, especially if the number of changes is limited.

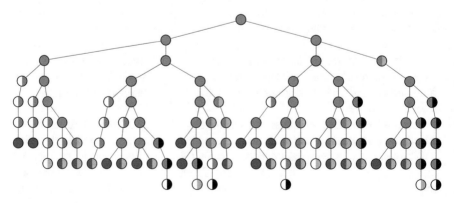

Fig. 2.1 Decoder tree for the $(1024, 512)$ polar code built using [61] and decoded with the nodes and operations of Table 2.1

Table 2.1 Decoder tree node types supported by the original Fast-SSC polar decoder [55]

Name	Color	Description
0R	White and gray	Left-half side is frozen
R1	Gray and black	Right-half side is all information
RSPC	Gray and yellow	Right-half side is an SPC code
0SPC	White and yellow	Left-half side is frozen, right-half side is an SPC code
Rep	Green	Repetition code, maximum length N_v of 16
RepSPC	Green and yellow	Concatenation of a repetition code on the left and an SPC code on the right, $N_v = 8$
01	Black and white	Fixed-length pattern $N_v = 4$ where the left-half side is frozen and the right-half side is all information
Rate-R	Gray	Mixed-rate node

2.2.2 Altered Polar Code Construction

In all SC-based decoders, the size of the decoder tree depends on the distribution of frozen and information bit locations in the code. Arıkan's original polar code construction only focuses on maximizing the reliability of the information bits. Several altered polar-like code constructions have been proposed in the literature [10, 25, 74] and their objective is to trade off error-correction performance for decoding complexity reduction by slightly changing the set of information bits, while keeping the code rate fixed. The main idea behind all the altered code constructions is to exchange the locations of a few frozen bits and information bits in order to get more bit patterns that are favorable in terms of decoding latency. In all cases, care must be taken in order to avoid using bit locations that are highly unreliable to transmit information bits.

The method in [25] first defines a small set of bit locations which contains the $n_s - h$ least reliable information bit locations along with the h most reliable frozen bit locations. Then, in order to keep the rate fixed, it performs an exhaustive search over all $\binom{n_s}{h}$ possible combinations of the n_s elements containing exactly h frozen bit locations and selects the combination that leads to the smallest decoding latency. In [10], the altered construction problem is formalized as a binary integer linear program. Consequently, it is shown that finding the polar code with the lowest decoding complexity under an error-correction performance constraint is an NP-hard problem. For this reason, a greedy approximation algorithm is presented which provides reasonable results at low complexity even for large code lengths. A similar greedy algorithm is presented in [74] for polar codes with more general code lengths of the form $N = l^n, l \geq 2$.

2.2.3 Proposed Altered Construction

The methods of [10, 74] only considered rate-0 and rate-1 nodes. As such, the results can not be directly applied to Fast-SSC decoding, where several additional types of special nodes exist. For this reason, in this work we follow the more general exhaustive search method of [25], augmented with a human-guided approach.

More specifically, bit-state alterations that would lead to smaller latency are identified by visual inspection of the decoder tree for the unaltered polar code. This list of bit locations is then passed to a program to be added to the bit locations considered by the technique described in [25]. Hence, two lists are composed: one that contains frozen bit locations proposed by the user as well as locations that were almost reliable enough to be used to carry information bits, and one that contains the information bit locations proposed by the user and the locations that barely made it into information bit locations.

The code alteration algorithm then proceeds by gradually calculating the decoding latency for all possible bit swap combinations. A constrained-size and ordered list of the combinations with the lowest decoding latency is kept. Once that list needs to be trimmed, only one entry per latency value is kept by simulating the error-correction performance of the altered code at an E_b/N_0 value of interest. The entry with the best FER is kept and the others with the same latency are removed from the list. That list containing the best candidates is further trimmed by removing all candidates that feature both a greater latency and worse error-correction performance compared to those of their predecessor. Similarly to the technique of [25], our proposed technique does not alter the code rate as the total number of information and frozen bits remains the same.

2.2.3.1 Human-Guided Criteria

The suggested bits to swap are selected to improve the latency and throughput. Thus, these bit swaps must eliminate constituent codes for which we do not have an efficient decoding algorithm and create ones for which we do. We classify the selection criteria under two categories: the bit swaps that transform frozen bit locations into information bit locations and bit swaps that do the opposite. The former increase the coding rate while the latter reduce it.

In addition to the node type definitions of Table 2.1, the below descriptions of criteria use the following types of subtrees or nodes:

- R1-01: subtree rooted in a R1 node with a 01 leaf node, may contain a chain of R1 nodes
- Rep1: subtree rooted in a R1 node with a leaf Rep node; in Sect. 2.3, that subtree is made into a node where the left-half side is a repetition code and the right-half side is all information
- R1-RepSPC: subtree rooted in a R1 node with a RepSPC leaf node, may contain a chain of R1 nodes

- Rep-Rep1: subtree where the rate-R node has a left-hand-side and right-hand-side nodes are Rep and Rep1 nodes, respectively
- 0-RepSPC: subtree rooted in a 0R node with a leaf RepSPC node; in Sect. 2.3, that subtree is made into a node where the left-half side is frozen and the right-half side is a RepSPC node

Dedicated hardware to efficiently decode Rep1 and 0RepSPC nodes are presented in Sect. 2.3.

From Frozen to Information Locations

1. Unfreezing the second bit of a 01 node that is part of a R1-01 subtree creates an RSPC node.
2. Changing an RepSPC into an RSPC node by adding the second, third and fifth bit locations.
3. Changing a RSPC node into a R1 node by changing the SPC code into a rate-1 code.

Criterion 1 is especially beneficial where the R1-01 subtree contains a chain of R1 nodes, e.g., Pattern 5 in Fig. 2.2. Similarly, Criterion 2 has a significant impact on R1-RepSPC subtrees containing a chain of R1 nodes, e.g., Pattern 3 in Fig. 2.2.

From Information to Frozen Locations

4. Changing a 0R-01 subtree into a Repetition node.
5. Freezing the only information bit location of a Rep node to change it into a rate-0 code.
6. A specialization of the above, changing a Rep-RepSPC subtree into a 0-RepSPC subtree by changing the left-hand-side Rep node into a rate-0 node.
7. Transforming a Rep-Rep1 subtree into a 0-RepSPC subtree by changing the left-hand-side Repetition code into a rate-0 code and by freezing the fifth bit location of the Rep1 subtree to change the rate-1 code into an SPC code.

Consider the decoder tree for a $(512, 376)$ polar code as illustrated in Fig. 2.2a, where some frozen bit patterns are circled in blue and numbered for reference. Its implementation results in a decoding latency of 106 clock cycles. That latency can be significantly reduced by freezing information bit locations or by transforming previously frozen locations into information bits.

Notably, five of the bit-swapping criteria—leading to latency reduction—described above are illustrated in Fig. 2.2a. The patterns numbered 1 and 2 are repetition nodes meeting the fourth criterion. Changing both into rate-0 nodes introduces two new 0R nodes. The patterns 3–6 are illustrations of the fourth, second, sixth and first criteria, respectively.

Figure 2.2b shows the resulting decoder tree after the alterations were made. The latency has been reduced from 106 to 82 clock cycles.

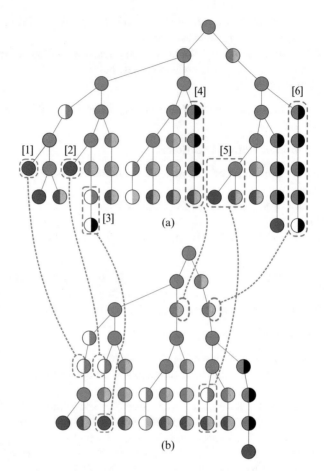

Fig. 2.2 Decoder trees for two different $(512, 376)$ polar codes, where (**a**) and (**b**) are before and after construction alteration, respectively

2.2.3.2   Example Results

Applying our proposed altered construction method, we were able to decrease the decoding latency of the $(1024, 512)$ polar code illustrated in Fig. 2.1 from 220 to 189 clock cycles, a 14% improvement, with 5 bit swaps. That increases the information throughput to 2.71 bits/CC, up from 2.33 bits/CC. The corresponding decoder tree is shown in Fig. 2.3.

The error-correction performance of the $(1024, 512)$ altered code is degraded as illustrated by the markerless black curves in Fig. 2.4. The loss amounts to less than $0.25\,\mathrm{dB}$ at a FER of 10^{-4}. For wireless applications, which are usually the target for codes of such lengths and rates, this represents the FER range of interest.

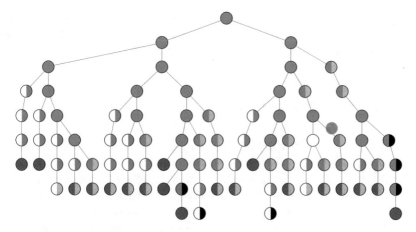

Fig. 2.3 Decoder tree for the altered $(1024, 512)$ polar code

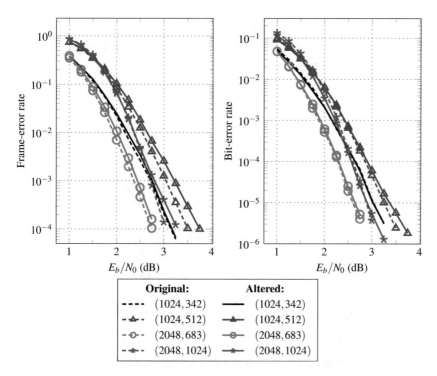

Fig. 2.4 Error-correction performance using BPSK over an AWGN channel of the altered codes compared to that of the original codes constructed using the Tal and Vardy method [61]

Figure 2.4 also shows the error-correction performance of three other polar codes altered using our proposed method. In the case of these other codes, the alterations have a negligible effect on error-correction performance.

2.3 New Constituent Decoders

Looking at the decoder tree of Fig. 2.3, it can be seen that some frozen bit patterns occur often. Adding support for more constituent codes to the Fast-SSC algorithm will result in a reduced latency and increased throughput under the constraint that the corresponding computation nodes do not significantly lengthens the critical path of a hardware implementation. As a result of an investigation, the constituent codes of Table 2.2 were added. Furthermore, post-place and route timing analysis showed that the maximum length N_v of a Repetition node could be increased from 16 to 32 without affecting the critical path.

The new decoder tree shown in Fig. 2.5 has a decoding latency of 165 clock cycles, a 13% reduction over the decoder tree of Fig. 2.3 decoded with the original Fast-SSC algorithm. Thus, the information throughput of that polar code has been improved to 3.103 bits/CC.

Table 2.2 New functions performed by the proposed decoder

Name	Color	Description
Rep1	Green and black	Repetition code on the left, Rate-1 code on the right, maximum length N_v of 8
0RepSPC	White and lilac	Rate-0 code on the left, RepSPC code on the right, $N_v = 16$
001	$\frac{3}{4}$ white and $\frac{1}{4}$ black	Rate-0 code on the left, 01 code on the right, $N_v = 8$

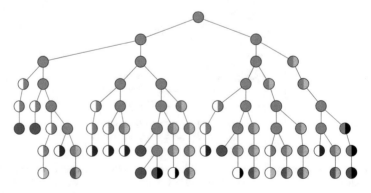

Fig. 2.5 Decoder tree for the altered polar code with the added nodes

Table 2.3 Frozen bit patterns decoded by leaf nodes

Name	Pattern
Rep	0001
	0000 0001
	0000 0000 0000 0001
	0000 0000 0000 0000 0000 0000 0000 0001
Rep1	0001 1111
0SPC	0000 0111
RepSPC	0001 0111
0RepSPC	0000 0000 0001 0111
01	0011
001	0000 0011

To summarize, Table 2.3 lists the frozen bit patterns that can be decoded by leaf nodes. It can be seen that the smallest possible leaf node has length $N_v = 4$ while our proposed decoder tree shown in Fig. 2.5 has a minimum length $N_v = 8$. In other words, Fig. 2.5 is representative of the patterns listed in Table 2.3 but not comprehensive.

2.4 Implementation

2.4.1 Quantization

Let Q_i be the total number of bits used to represent LLRs internally, Q_c be the total number of bits to represent channel LLRs, and Q_f be the number of bits among Q_i or Q_c used to represent the fractional part of any LLR. It was found through simulations that using $Q_i.Q_c.Q_f = 6.5.1$ quantization led to an error-correction performance very close to that of the floating-point number representation as can be seen in Fig. 2.6.

2.4.2 Rep1 Node

The Rep1 node decodes Rep1 codes—the concatenation of a repetition code and a rate-1 code—of length $N_v = 8$. Its bit-estimate vector β_0^7 is calculated using operations described in the previous sections. However, instead of performing the required operations sequentially, the dedicated hardware preemptively calculates intermediate soft values.

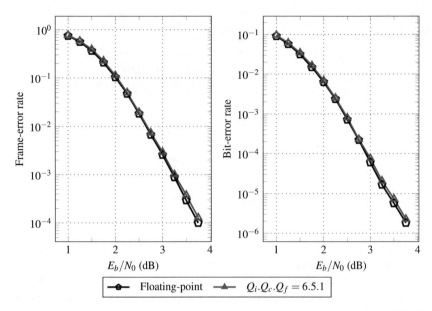

Fig. 2.6 Impact of quantization on the error-correction performance of the proposed $(1024, 512)$ polar code

Fig. 2.7 Architecture of the Rep1 Node

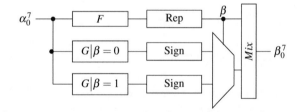

Figure 2.7 shows the architecture of the Rep1 node. It can be seen that there are two G blocks. One preemptively calculates soft values assuming that the Rep block will output $\beta = 0$ and the other for $\beta = 1$. The Rep block provides a single bit estimate corresponding to the information bit the repetition code of length $N_v = 4$ it is decoding. The outputs of the G blocks go through a Sign block to generate hard decisions. The correct hard decision vector is then selected using the output of the Rep block. Finally, the bit estimate vector β_0^7 is built. The highest part, β_4^7, is always comprised of the multiplexer output. The lowest part, β_0^3, is either a copy of same output or its binary negation. The negated version is selected when the output of the Rep block is 1.

Calculations are carried out in one clock cycle. The output of the F, G and Rep blocks are not stored in memory. Only the final result, the bit-estimate vector β_0^7, is stored in memory.

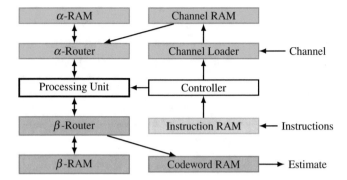

Fig. 2.8 High-level architecture of the decoder

2.4.3   High-Level Architecture

The high-level architecture of the decoder is presented in Fig. 2.8. Instructions representing the polar decoding operations to be performed are loaded before decoding starts. When the decoder is started, the controller signals the channel loader to start storing channel LLRs, 32 LLRs (160 bits) per clock cycle, into the channel RAM. The controller then starts to execute functions on the processing unit. The processing unit reads LLRs from the Channel or α-RAM and writes LLRs to the α-RAM. It reads or writes hard decisions to the β-RAM. The last *Combine* operation writes the estimated codeword into the Codeword RAM, a memory accessible from outside the decoder.

The decoder is complete with all input and output buffers to accommodate loading a new frame and reading an estimated codeword while a frame is being decoded. The required memory could be made smaller if the nominal throughput required is lower. The loading or outputting of a full frame takes fewer clock cycles than the actual decoding, we have a pipelined operation; under normal operation, the decoder should not be slowed down by the Input/Output (I/O) operations.

2.4.4   Processing Unit or Processor

The core of the decoder is the processing unit illustrated in Fig. 2.9 and based on the Fast-SSC implementation of [55]. Thus, the processing unit features all the modules required to implement the nodes and operations listed in Table 2.1 and described in Sect. 2.3. Notably, the 01 and RepSPC blocks connected to the G block implement the 001 and 0RepSPC nodes, respectively, where the all-zero vector input is selected at the multiplexer m_0. The critical path of the decoder corresponds to the 0RepSPC node i.e. goes through G, RepSPC, the multiplexer m_3, *Combine* and the multiplexer m_2. It is slightly longer than that of [55].

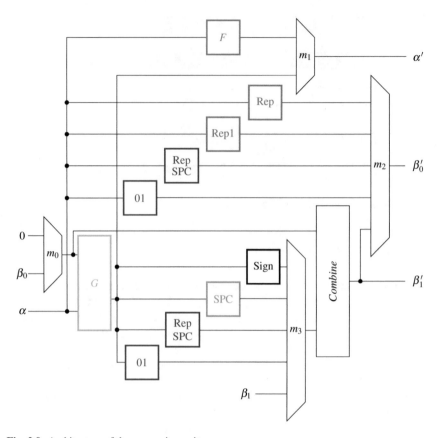

Fig. 2.9 Architecture of the processing unit

2.5 Results

2.5.1 *Verification Methodology*

A software model was used to generate random codewords for transmission using Binary Phase-Shift Keying (BPSK) over an Additive White Gaussian-Noise (AWGN) channel. The functionality of the designs was verified both at the Register-Transfer Level (RTL) and at the post-place and route level through simulations. Finally, the same frames were also decoded on an FPGA using an FPGA-in-the-loop setup. For all E_b/N_0 values, a minimum of 100 frames in errors were simulated.

2.5.2 Comparison with State-of-the-Art Decoders

In this section, post-fitting results are presented for the Altera Stratix IV EP4SGX530KH40C2 FPGA. All results are worst-case using the slow 900 mV 85 °C timing model. Table 2.4 shows the results for two rate-flexible implementations for polar codes of length 1024 and 2048, respectively. The decoder of [48] is also included for comparison.

Looking at the results for our proposed decoders, it can be observed that the number of Look-Up Tables (LUTs) and registers required are very similar for both code lengths. However, the RAM usage differs significantly where decoding a longer code requires more memory as expected. Timing reports show that the critical path corresponds to the 0RepSPC node.

Table 2.4 also compares the proposed decoders against the decoder of [48] as well as the original Fast-SSC implementation [55]. The latter was resynthesized so that the decoder only has to accommodate polar codes of length $N = 1024$ or $N = 2048$ and is marked with an asterisk (*) in Table 2.4.

Our work requires at most a 1.4% increase in used LUTs compared to [55]. The difference in registers can be mostly attributed to register duplication, a measure taken by the fitter to shorten the critical path to meet the requested clock frequency. The Static Random-Access Memory (SRAM) usage was also increased by 2.3%.

Table 2.5 shows the latency and information throughput of the decoders of Table 2.4 when decoding low-rate moderate-length polar codes. It also shows the effect of using a polar codes with altered constructions—as described in Sect. 2.2— with all Fast-SSC-based decoders. For both [55]* and our work, the results listed as 'altered codes' have the same resource usage and clock frequency as listed in Table 2.4 since these decoders can decode any polar code of length $N = 1024$ or $N = 2048$ by changing the code description in memory.

Applying the proposed altered construction alone, Table 2.5 shows that decoding these altered codes with the original decoders of [55] results in a 14–21% latency reduction and a 16–27% throughput improvement. From the same table, it can be seen that decoding the unaltered codes with the updated hardware decoder integrating the proposed new constituent decoders, the latency is reduced by 4–10% and the throughput is improved by 4–10%.

Table 2.4 Post-fitting results for rate-flexible decoders for moderate-length polar codes

Implementation	N	LUTs	Regs.	RAM (kbits)	f (MHz)
[48]	1024	1940	748	7.1	239
[55]*	1024	23 020	1024	42.8	103
	2048	23 319	5923	60.9	103
this work	1024	23 353	5814	43.8	103
	2048	23 331	5923	61.2	103

Table 2.5 Latency and information throughput comparison for low-rate moderate-length polar codes

Implementation	Code (N, k)	Latency		Info. T/P (Mbps)
		(CCs)	(μs)	
[48]	(1024, 342)	2185	9.14	37
	(1024, 512)	2185	9.14	56
[55]*	(1024, 342)	201	1.95	175
	(1024, 512)	220	2.14	240
	(2048, 683)	366	3.55	192
	(2048, 1024)	389	3.78	271
Altered codes	(1024, 342)	173	1.68	204
	(1024, 512)	186	1.81	284
	(2048, 683)	289	2.81	243
	(2048, 1024)	336	3.26	314
This work	(1024, 342)	193	1.87	183
	(1024, 512)	204	1.98	259
	(2048, 683)	334	3.24	211
	(2048, 1024)	367	3.56	287
Altered codes	(1024, 342)	157	1.52	224
	(1024, 512)	165	1.60	320
	(2048, 683)	274	2.66	257
	(2048, 1024)	308	2.99	342

Combining the contribution of both the altered construction method and the new dedicated constituent decoders, the proposed work achieves the best latency among all compared decoders. For the polar codes of length $N = 1024$, the throughput is 5.7–6.1 times greater than that of the two-phase decoder of [48]. Finally, the latency is reduced by 22–28% and the throughput is increased by 26–34% over the Fast-SSC decoders of [55].

Table 2.6 presents a comparison of this work against the state-of-the-art ASIC implementations. Our ASIC results are for the 65 nm CMOS GP technology from TSMC and are obtained with Cadence RTL Compiler. Only registers were used for memory due to the lack of access to an SRAM compiler. Normalized results for the decoders from the literature are also provided. For consistency, only results for a (1024, 512) polar code are compared to match what was done in the other works. It should be noted that [49] provides measurement results.

From Table 2.6, it can be seen that both implementations of our proposed decoder—at different supply voltages—are 46 and 42% the size of the BP decoder [49] and the combinational decoder [14], respectively, when all areas are normalized to 65 nm technology. Our work has two orders of magnitude lower latency than the BP decoder of [49], and two to five times lower latency than [72]. The latency of the proposed design is 1.05 times and 0.7 times that of [14], when operating at 400 and 600 MHz, respectively. The BP decoder [49] employs early termination

Table 2.6 Comparison of state-of-the-art ASIC decoders decoding a (1024, 512) polar code

	This work		[49][a]	[14]	[72]
Algorithm	Fast-SSC		BP	SC	2-bit SC
Technology	65 nm		65 nm	90 nm	45 nm
Supply (V)	0.8	1.0	1.0	1.3	N/A
Oper. temp. (°C)	25	25	$\approx$ 25	N/A	N/A
Area (mm^2)	0.69	0.69	1.48	3.21	N/A
Area @65nm (mm^2)	0.69	0.69	1.48	1.68	0.4
Frequency (MHz)	400	600	300	2.5	750
Latency (μs)	0.41	0.27	50	0.39	1.02
Info. T/P (Gbps)	1.24	1.86	2.4 @ 4dB	1.28	0.5
Sust. Info. T/P (Gbps)	1.24	1.86	1.0	1.28	0.5
Area Eff. (Gbps/mm^2)	1.8	2.7	1.6 @ 4dB	0.4	N/A
Power (mW)	96	215	478	191	N/A
Energy (pJ/bit)	77	115	203 @ 4dB	149	N/A

[a]Measurement results

and its throughput at $E_b/N_0 = 4$ dB is the fastest followed by our proposed design. Since the area reported in [49] excludes the memory necessary to buffer additional received vectors to sustain the variable decoding latency due to early termination, we also report the sustained throughput for that decoder. The sustained throughput is 1.0 Gbps as a maximum of 15 iterations is required for the BP decoder to match the error-correction performance of the SC-based decoders. Comparing the information throughput of all decoders—using the best-case values for BP—it can be seen that the area efficiency of our decoder is the greatest. Lastly, the power consumption estimations indicate that our decoders are more energy efficient than the BP decoder of [49]. Our proposed decoders are also more energy efficient than that of [14]. However, due to the difference in implementation technology, the results of this latter comparison could change if [14] were to be implemented in 65 nm.

2.6 Conclusion

In this chapter, we showed how the original Fast-SSC algorithm implementation could be improved by adding dedicated decoders for three new types of constituent codes frequently appearing in low-rate codes. We also used polar code construction alterations to significantly reduce the latency and increase the throughput of a Fast-SSC decoder at the cost of a small error-correction performance loss. Rate-flexible decoders for polar codes of lengths 1024 and 2048 were implemented on an FPGA. Four low-rate polar codes with competitive error-correction performance were proposed. Their resulting latency and throughput represent a 22–28% reduction and a 26–34% improvement over the previous work, respectively. The information throughput was shown to be 224, 320, 257, and 342 Mbps at approximately

100 MHz on the Altera Stratix IV FPGAs for the $(1024, 342)$, $(1024, 512)$, $(2048, 683)$ and $(2048, 1024)$ polar codes, respectively. On 65 nm ASIC CMOS technology, the proposed decoder for a $(1024, 512)$ polar code was shown to compare favorably against the state-of-the-art ASIC decoders. With a clock frequency of 400 MHz and a supply voltage of 0.8 V, it has a latency of 0.41 μs and an area efficiency of 1.8 Gbps/mm^2 for an energy efficiency of 77 pJ/info. bit. At 600 MHz with a supply of 1 V, the latency is reduced to 0.27 μs and the area efficiency increased to 2.7 Gbps/mm^2 at 115 pJ/info. bit.

Chapter 3
Low-Latency Software Polar Decoders

Abstract The low-complexity encoding and decoding algorithms render polar codes attractive for use in SDR applications where computational resources are limited. In this chapter, we present low-latency software polar decoders that exploit modern processor capabilities. We show how adapting the algorithm at various levels can lead to significant improvements in latency and throughput, yielding polar decoders that are suitable for high-performance SDR applications on modern desktop processors and embedded-platform processors. These proposed decoders have an order of magnitude lower latency and memory footprint compared to state-of-the-art decoders, while maintaining comparable throughput. In addition, we present strategies and results for implementing polar decoders on graphical processing units. Finally, we show that the energy efficiency of the proposed decoders is comparable to state-of-the-art software polar decoders.

3.1 Introduction

In SDR applications, researchers and engineers have yet to fully harness the error-correction capability of modern codes due to their high computational complexity. Many are still using classical codes [13, 63] as implementing low-latency high-throughput—exceeding 10 Mbps of information throughput—software decoders for turbo or LDPC codes is very challenging. The irregular data access patterns featured in decoders of modern error-correction codes make efficient use of SIMD extensions present on today's Central Processing Units (CPUs) difficult. To overcome this difficulty and still achieve a good throughput, software decoders resorting to inter-frame parallelism (decoding multiple independent frames at the same time) are often proposed [30, 66, 69]. Inter-frame parallelism comes at the cost of higher latency, as many frames have to be buffered before decoding can be started. Even with a split layer approach to LDPC decoding where intra-frame parallelism can be applied, the latency remains high at multiple milliseconds on a recent desktop processor [23]. This work presents software polar decoders that enable SDR systems to utilize powerful *and* fast error-correction.

© Springer International Publishing AG 2017
P. Giard et al., *High-Speed Decoders for Polar Codes*,
DOI 10.1007/978-3-319-59782-9_3

Polar codes provably achieve the symmetric capacity of memoryless channels [7]. Moreover they are well suited for software implementation, due to regular memory access patterns, on both x86 and embedded processors [18, 31, 32]. To achieve higher throughput and lower latency on processors, software polar decoders can also exploit SIMD vector extensions present on today's CPUs. Vectorization can be performed intra-frame [18] or inter-frame [31, 32], with the former having lower decoding latency as it does not require multiple frames to start decoding.

In this work, we explore intra-frame vectorized polar decoders. We propose architectures and optimization strategies that lead to the implementation of high-performance software polar decoders tailored to different processor architectures with decoding latency of 26 µs for a (32 768, 29 492) polar code, an order of magnitude performance improvement compared to that of our earlier work [18]. We start Sect. 3.2 by presenting two different software decoder architectures with varying degrees of specialization. Implementation and results on an embedded processor are discussed in Sect. 3.3. We also adapt the decoder to suit Graphical Processing Units (GPUs), an interesting target for applications where many hundreds of frames have to be decoded simultaneously, and present the results in Sect. 3.4. Finally, Sect. 3.5 compares the energy consumption of the different decoders and Sect. 3.7 concludes this chapter.

This chapter builds upon the work we published in [18] and co-authored in [54]. It provides additional details on the approach as well as more experimental results for modern desktop processors. Both floating- and fixed-point implementations for the final desktop CPU version—the unrolled decoder—were further optimized leading to an information throughput of up to 1.4 Gbps. It also adds results for the adaptation of our strategies to an embedded processor leading to a throughput and latency of up to 2.25 and 36 times better, respectively, compared to that of the state-of-the-art software implementation. Compared to the state of the art, both the desktop and embedded processor implementations are shown to have one to two orders of magnitude smaller memory footprint. Lastly, strategies and results for implementing polar decoders on a GPU are presented for the first time.

3.2 Implementation on x86 Processors

In this section we present two different versions of the decoder in terms of increasing design specialization for software; whereas the first version—the instruction-based decoder—takes advantage of the processor architecture it remains configurable at run time and the second one—the unrolled decoder—presents a fully unrolled, branchless decoder fully exploiting SIMD vectorization. In the second version of the decoder, compile-time specialization plays a significant role in the performance improvements. Performance is evaluated for both the instruction-based and unrolled decoders.

It should be noted that, contrary to what is common in hardware implementations e.g. [33, 55], natural indexing is used for all software decoder implementations.

While bit-reversed indexing is well-suited for hardware decoders, SIMD instructions operate on independent vectors, not adjacent values within a vector. Using bit-reversed indexing would have mandated data shuffling operations before any vectorized operation is performed.

Both versions, instruction-based decoders and unrolled decoders, use the following functions from the Fast-SSC algorithm [55]: *F*, *G*, *G0R*, *Combine*, *C0R*, Repetition, 0SPC, RSPC, RepSPC and P01. An Info function implementing equation (1.5) is also added.

Methodology for the Experimental Results

We discuss throughput in information bits per second as well as latency. Our software was compiled using the C++ compiler from GCC 4.9 using the flags "`-march=native -funroll-loops -Ofast`". Additionally, auto-vectorization is always kept enabled. The decoders are inserted in a digital communication chain to measure their speed and to ensure that optimizations, including those introduced by `-Ofast`, do not affect error-correction performance. In the simulations, we use BPSK over an AWGN channel with random codewords.

The throughput is calculated using the time required to decode a frame averaged over 10 runs of 50 000 and 10 000 frames each for the $N = 2048$ and the $N > 2048$ codes, respectively. The time required to decode a frame, or latency, also includes the time required to copy a frame to decoder memory and copy back the estimated codeword. Time is measured using the high precision clock provided by the Boost Chrono library.

In this work we focus on decoders running on one processor core only since the targeted application is SDR. Typically, an SDR system cannot afford to dedicate more than a single core to error correction as it has to perform other functions simultaneously. For example, in SDR implementations of Long-Term Evolution (LTE) receivers, the orthogonal frequency-division multiplexing (OFDM) demodulation alone is approximately an order of magnitude more computationally demanding than the error-correction decoder [11, 13, 63].

3.2.1 Instruction-Based Decoder

The Fast-SSC decoder implemented on a FPGA in [55] closely resembles a CPU with wide SIMD vector units and wide data buses. Therefore, it was natural to use a similar design for a software decoder—Fast-SSC instructions are parsed from a text file—leveraging SIMD instructions. This section describes how the algorithm was adapted for a software implementation. As fixed-point arithmetic can be used, the effect of quantization is shown.

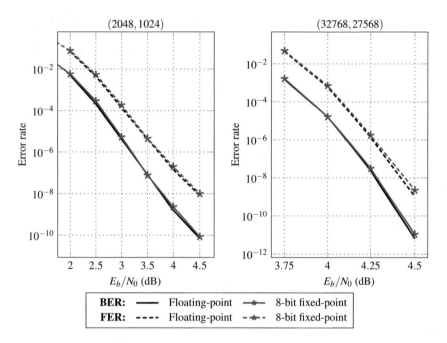

Fig. 3.1 Effect of quantization on error-correction performance

3.2.1.1 Using Fixed-Point Numbers

On processors, fixed-point numbers are represented with at least 8 bits. As illustrated in Fig. 3.1, using 8 bits of quantization for LLRs results in a negligible degradation of error-correction performance over a floating-point representation. At a FER of 10^{-8} the performance loss compared to a floating-point implementation is less than 0.025 dB for the $(32\,768, 27\,568)$ polar code. With custom hardware, it was shown in [55] that 6 bits are sufficient for that polar code. It should be noted that in Fast-SSC decoding, only the G function adds to the amplitude of LLRs and it is carried out with saturating adders.

With instructions that can work on registers of packed 8-bit integers, the SIMD extensions available on most general-purpose x86 and ARM processors are a good fit to implement a polar decoder.

3.2.1.2 Vectorizing the Decoding of Constituent Codes

On x86-64 processors, the vector instructions added with Streaming SIMD Extensions (SSE) support logic and arithmetic operations on vectors containing either four single-precision floating-point numbers or 16 8-bit integers. Additionally, x86-64 processors with Advanced Vector eXtensions (AVX) instructions can operate on data sets of twice that size. Below are the operations benefiting the most from explicit vectorization.

F: the *f* operation (1.1) is often executed on large vectors of LLRs to prepare values for other processing nodes. The min() operation and the sign calculation and assignment are all vectorized.

G and *G0R*: the *g* operation is also frequently executed on large vectors. Both possibilities, the sum and the difference, of (1.2) are calculated and are blended together with a mask to build the result. The *G0R* operation replaces the G operation when the left hand side of the tree is the all-zero vector.

Combine and *C0R*: the Combine operation combines two estimated bit-vectors using an XOR operation in a vectorized manner. The *C0R* operation is to Combine what *G0R* is to G.

SPC decoding: locating the LLR with the minimum magnitude is accelerated using SIMD instructions.

3.2.1.3 Data Representation

For the decoders using floating-point numbers, the representation of β is changed to accelerate the execution of the *g* operation on large vectors. Thus, when floating-point LLRs are used, $\beta_l[i] \in \{+1, -1\}$ instead of $\{0, 1\}$. As a result, (1.2) can be rewritten as

$$g(\alpha_v[i], \alpha_v[i + {}^{N_v}/_2], \beta_l[i]) = \alpha_v[i] * \beta_l[i] + \alpha_v[i + {}^{N_v}/_2].$$

This removes the conditional assignment and turns $g()$ into a multiply-accumulate operation, which can be performed efficiently in a vectorized manner on modern CPUs. For integer LLRs, multiplications cannot be carried out on 8-bit integers. Thus, both possibilities of (1.2) are calculated and are blended together with a mask to build the result. The Combine operation is modified accordingly for the floating-point decoder and is computed using a multiplication with $\beta_l[i] \in \{+1, -1\}$.

3.2.1.4 Architecture-Specific Optimizations

The decoders take advantage of the Supplemental Streaming SIMD Extensions (SSSE) 3, SSE 4.1 and AVX instructions when available. Notably, the `sign` and `abs` instructions from SSSE 3 and the `blendv` instruction from SSE 4.1 are used. AVX, with instructions operating on vectors of 256 bits instead of the 128 bits, is only used for the floating-point implementation since it does not support integer operations. Data was aligned to the 128 (SSE) or 256-bit (AVX) boundaries for faster accesses.

Table 3.1 Decoding polar codes with the instruction-based decoder

Code (N, k)	Implementation	Info T/P (Mbps)	Latency (μs)
(2048, 1024)	Float	20.8	49
	SIMD-Float	75.6	14
	SIMD-int8	121.7	8
(2048, 1707)	Float	41.5	41
	SIMD-Float	173.9	10
	SIMD-int8	209.9	8
(32 768, 27 568)	Float	32.4	825
	SIMD-Float	124.3	222
	SIMD-int8	175.1	157
(32 768, 29 492)	Float	40.8	723
	SIMD-Float	160.1	184
	SIMD-int8	198.6	149

3.2.1.5   Implementation Comparison

Here we compare the performance of three implementations. First, a non-explicitly vectorized version[1] using floating-point numbers. Second an explicitly vectorized version using floating-point numbers. Third, the explicitly vectorized version using a fixed-point number representation. In Table 3.1, they are denoted as Float, SIMD-Float and SIMD-int8 respectively.

Results for decoders using the floating-point number representation are included as the efficient implementation makes the resulting throughput high enough for some applications. The decoders ran on a single core of an Intel Core i7-4770S clocked at 3.1 GHz with Turbo disabled.

Comparing the throughput and latency of the Float and SIMD-Float implementations in Table 3.1 confirms the benefits of explicit vectorization in this decoder. The performance of the SIMD-Float implementation is only 21–38% slower than the SIMD-int8 implementation. This is not a surprising result considering that the SIMD-Float implementation uses the AVX instructions operating on vectors of 256 bits while the SIMD-int8 version is limited to vectors of 128 bits. Table 3.1 also shows that vectorized implementations have 3.6–5.8 times lower latency than the floating-point decoder.

---

[1] As stated above, compiler auto-vectorization is always kept enabled.

3.2.2 Unrolled Decoder

The goal of this design is to increase vectorization and inlining and reduce branches in the resulting decoder by maximizing the information specified at compile time. It also gets rid of the indirections that were required to get good performance out of the instruction-based decoder.

3.2.2.1 Generating an Unrolled Decoder

The polar codes decoded by the instruction-based decoders presented in Sect. 3.2.1 can be specified at run time. This flexibility comes at the cost of increased branches in the code due to conditionals, indirections and loops. Creating a decoder dedicated to only one polar code enables the generation of a branchless fully-unrolled decoder. In other words, knowing in advance the dimensions of the polar code and the frozen bit locations removes the need for most of the control logic and eliminates branches there.

A tool was built to generate a list of function calls corresponding to the decoder tree traversal. It was first described in [54] and has been significantly improved since its initial publication notably to add support for other node types as well as to add support for GPU code generation. Listing 1 shows an example decoder that corresponds to the $(8, 5)$ polar code whose dataflow graph is shown in Fig. 3.2. For brevity and clarity, in Fig. 3.2b, I corresponds to the Info function.

Listing 1 Unrolled $(8, 5)$ Fast-SSC Decoder

$F{<}8{>}(\alpha_c, \alpha_1)$;
$G0R{<}4{>}(\alpha_1, \alpha_2)$;
$Info{<}2{>}(\alpha_2, \beta_1)$;
$C0R{<}4{>}(\beta_1, \beta_2)$;
$G{<}8{>}(\alpha_c, \alpha_2, \beta_2)$;
$SPC{<}4{>}(\alpha_2, \beta_3)$;
$Combine{<}8{>}(\beta_2, \beta_3, \beta_c)$;

Fig. 3.2 Dataflow graph of a $(8, 5)$ polar decoder.
(a) Messages. **(b)** Operations

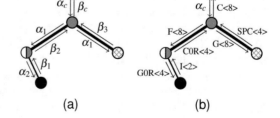

(a)   (b)

3.2.2.2 Eliminating Superfluous Operations on β-Values

Every non-leaf node in the decoder performs the combine operation (1.3), rendering it the most common operation. In (1.3), half the β values are copied unchanged to β_v. One method to significantly reduce decoding latency is to eliminate those superfluous copy operations by choosing an appropriate layout for β values in memory: Only N β values are stored in a contiguous array aligned to the SIMD vector size. When a combine operation is performed, only those values corresponding to β_l will be updated. Since the stage sizes are all powers of two, stages of sizes equal to or larger than the SIMD vector size will be implicitly aligned so that operations on them are vectorized.

3.2.2.3 Improved Layout of the α-Memory

Unlike in the case of β values, the operations producing α values, f and g operations, do not copy data unchanged. Therefore, it is important to maximize the number of vectorized operations to increase decoding speed. To this end, contiguous memory is allocated for the $\log_2 N$ stages of the decoder. The overall memory and each stage is aligned to 16 or 32-byte boundaries when SSE or AVX instructions are used, respectively. As such, it becomes possible to also vectorize stages smaller than the SIMD vector size. The memory overhead due to not tightly packing the stages of α memory is negligible. As an example, for an $N = 32\,768$ floating-point polar decoder using AVX instructions, the size of the α memory required by the proposed scheme is 262 208 bytes, including a 60-byte overhead.

3.2.2.4 Compile-Time Specialization

Since the sizes of the constituent codes are known at compile time, they are provided as template parameters to the functions as illustrated in Listing 1. Each function has two or three implementations. One is for stages smaller than the SIMD vector width where vectorization is not possible or straightforward. A second one is for stages that are equal or wider than the largest vectorization instruction set available. Finally, a third one provides SSE vectorization in an AVX or AVX2 decoder for stages that can be vectorized by the former, but are too small to be vectorized using AVX or AVX2. The last specialization was noted to improve decoding speed in spite of the switch between the two SIMD extension types.

Furthermore, since the bounds of loops are compile-time constants, the compiler is able to unroll loops where it sees fit, eliminating the remaining branches in the decoder unless they help in increasing speed by resulting in a smaller executable.

3.2.2.5 Architecture-Specific Optimizations

First, the decoder was updated to take advantage of AVX2 instructions when available. These new instructions benefit the fixed-point implementation as they allow simultaneous operations on 32 8-bit integers.

Second, the implementation of some nodes were hand optimized to better take advantage of the processor architecture. For example, the SPC node was mostly rewritten. Listing 2 shows a small but critical subsection of the SPC node calculations where the index within a SIMD vector corresponding to the specified value is returned. The reduction operation required by the Repetition node has also been optimized manually.

Third, for the floating-point implementation, β was changed to be in $\{+0, -0\}$ instead of $\{+1, -1\}$. In the floating-point representation [2], the most significant bit only carries the information about the sign. Flipping this bit effectively changes the sign of the number. By changing the mapping for β, multiplications are replaced by faster bitwise XOR operations. Similarly, for the 8-bit fixed-point implementation, β was changed to be in $\{0, -128\}$ to reduce the complexity of the Info and G functions.

Listings 3 and 4 show the resulting G functions for both the floating-point and fixed-point implementations as examples illustrating bottom-up optimizations used in our decoders.

Listing 2 Finding the index of a given value in a vector

```
std::uint32_t findIdx(α* x, α x_min) {
    __mm256 minVec = _mm256_broadcastb_epi8(x_min);
    __mm256 mask = _mm256_cmpeq_epi8(minVec, x);
    std::uint32_t mvMask = _mm256_movemask_epi8(mask);
    return __tzcnt_u32(mvMask);
}
```

Listing 3 Vectorized floating-point G function (g operation)

```
template<unsigned int N_v>
void G(α* α_in, α* α_out, β* β_in) {
    for (unsigned int i = 0; i < N_v/2; i += 8) {
        __m256 α_l = _mm256_load_ps(α_in + i);
        __m256 α_r = _mm256_load_ps(α_in + i + N_v/2);
        __m256 β_v = _mm256_load_ps(β_in + i);
        __m256 α'_l = _mm256_xor_ps(β_v, α_l);
        __m256 α_v = _mm256_add_ps(α_r, α'_l);
        __mm256_store_ps(α_out + i, α_v);
    }
}
```

Listing 4 Vectorized 8-bit fixed-point G function (g operation)

```
static const __m256i ONE = _mm256_set1_epi8(1);
static const __m256i M127 = _mm256_set1_epi8(−127);

template<unsigned int Nᵥ>
void G(α* αin, α* αout, β* βin) {
    for (unsigned int i = 0; i < Nᵥ/2; i += 32) {
        __m256i αl = _mm256_load_si256(αin + i);
        __m256i αr = _mm256_load_si256(αin + i + Nᵥ/2);
        __m256i βv = _mm256_load_si256(βin + i);
        __m256i β′v = _mm256_or_si256(βv, ONE);
        __m256i α′l = _mm256_sign_epi8(αl, β′v);
        __m256i αv = _mm256_add_ps(αr, α′l);
        __m256i α′v = _mm256_max_epi8(M127, αv);
        _mm256_store_si256(αout + i, α′v);
    }
}
```

3.2.2.6 Memory Footprint

The memory footprint is considered an important constraint for software applications. Our proposed implementations use two contiguous memory blocks that correspond to the α and β values, respectively. The size of the β-memory is

$$M_\beta = NW_\beta, \tag{3.1}$$

where N is the frame length, W_β is the number of bits used to store a β value and M_β is in bits.

The size of the α-memory can be expressed as

$$M_\alpha = \left\lceil (2N - 1) + A\log_2 A - \left(\sum_{i=0}^{\log_2(A)-1} 2^i \right) \right\rceil W_\alpha, \tag{3.2}$$

where N is the frame length, W_α is the number of bits used to store an α value, A is the number of α values per SIMD vector and M_α is in bits. Note that the expression of M_α contains the expression for the overhead $M_{\alpha\text{OH}}$ due to tightly packing the α values as described in Sect. 3.2.2.3:

$$M_{\alpha\text{OH}} = \left\lceil A\log_2 A - \left(\sum_{i=0}^{\log_2(A)-1} 2^i \right) \right\rceil W_\alpha. \tag{3.3}$$

The memory footprint can thus be expressed as

$$M_{total} = M_\beta + M_\alpha$$

$$= NW_\beta + \left[(2N - 1) + A \log_2 A - \left(\sum_{i=0}^{\log_2(A)-1} 2^i \right) \right] W_\alpha. \tag{3.4}$$

The memory footprint in kilobytes can be approximated with

$$M_{total\ (kbytes)} \approx \frac{N(W_\beta + 2W_\alpha)}{8000}. \tag{3.5}$$

3.2.2.7 Implementation Comparison

We first compare the SIMD-float results for this implementation—the unrolled decoder—with those from Sect. 3.2.1—the instruction-based decoder. Then we show SIMD-int8 results and compare them with that of the software decoder of Le Gal et al. [32]. As in the previous sections, the results are for an Intel Core i7-4770S running at 3.1 GHz when Turbo is disabled and at up to 3.9 GHz otherwise. The decoders were limited to a single CPU core.

Table 3.2 shows the impact of the optimizations introduced in the unrolled version on the SIMD-float implementations. It resulted in the unrolled decoders being two to three times faster than the flexible, instruction-based, ones. Comparing Tables 3.1 and 3.2 shows an improvement factor from 3.3 to 5.7 for the SIMD-int8 implementations. It should be noted that some of the improvements introduced in the unrolled decoders could be backported to the instruction-based decoders, and is considered for future work.

Compared to the software polar decoders of [32], Table 3.3 shows that our throughput is lower for short frames but can be comparable for long frames. However, latency is an order of magnitude lower for all code lengths. This is to be expected as the decoders of [32] do inter-frame parallelism i.e. parallelize the decoding of independent frames while we parallelize the decoding of a frame. The memory footprint of our decoder is shown to be approximately 24 times lower than

Table 3.2 Decoding polar codes with floating-point precision using SIMD, comparing the instruction-based decoder (ID) with the unrolled decoder (UD)

Code (N, k)	Info T/P (Mbps)		Latency (µs)	
	ID	UD	ID	UD
(2048, 1024)	75.6	229.8	14	4
(2048, 1707)	173.9	492.2	10	3
(32 768, 27 568)	124.3	271.3	222	102
(32 768, 29 492)	160.1	315.1	184	94

Table 3.3 Comparison of the proposed software decoder with that of [32]

Decoder	Target	L3 Cache	f (GHz)	Code (N, k)	Mem. footprint (kbytes)	Info T/P (Mbps)	Latency (μs)
[32]*	Intel Core i7-4960HQ	6 MB	3.6+	(2048, 1024)	144	1320	25
				(2048, 1707)	144	2172	26
				(32 768, 27 568)	2304	1232	714
				(32 768, 29 492)	2304	1557	605
This work	Intel Core i7-4770S	8 MB	3.1	(2048, 1024)	6	398	3
				(2048, 1707)	6	1041	2
				(32 768, 27 568)	98	886	31
				(32 768, 29 492)	98	1131	26
This work*	Intel Core i7-4770S	8 MB	3.1+	(2048, 1024)	6	502	2
				(2048, 1707)	6	1293	1
				(32 768, 27 568)	98	1104	25
				(32 768, 29 492)	98	1412	21

*Results with Turbo enabled

Table 3.4 Effect of unrolling and algorithm choice on decoding speed of the (2048, 1707) code on the Intel Core i7-4770S

Decoder	Info T/P (Mbps)	Latency (μs)
ID Fast-SSC	210	8.1
UD SC	363	4.7
UD Fast-SSC	1041	1.6

that of [32]. The results in [32] were presented with Turbo frequency boost enabled; therefore we present two sets of results for our proposed decoder: one with Turbo enabled, indicated by the asterisk (*) and the 3.1+ GHz frequency in the table, and one with Turbo disabled. The results with Turbo disabled are more indicative of a full SDR system as all CPU cores will be fully utilized, not leaving any thermal headroom to increase the frequency. The maximum Turbo frequencies are 3.8 and 3.9 GHz for the i7-4960HQ and i7-4770S CPUs, respectively.

Looking at the first two, or last two rows of Table 3.2, it can be seen that for a fixed code length, the decoding latency is smaller for higher code rates. The tendency of decoding latency to decrease with increasing code rate and length was first discussed in [57]. It was noted that higher rate codes resulted in SSC decoder trees with fewer nodes and, therefore, lower latency. Increasing the code length was observed to have a similar, but lesser, effect. However, once the code becomes sufficiently long, the limited memory bandwidth and number of processing resources form bottlenecks that negate the speed gains.

The effects of unrolling and using the Fast-SSC algorithm instead of SC are illustrated in Table 3.4. It can be observed that unrolling the Fast-SSC decoder results in a five time decrease in latency. Using the Fast-SSC instead of SC decoding algorithm decreased the latency of the unrolled decoder by three times.

3.3 Implementation on Embedded Processors

Many of the current embedded processors used in SDR applications also offer SIMD extensions, e.g. NEON for ARM processors. All the strategies used to develop an efficient x86 implementation can be applied to the ARM architecture with changes to accommodate differences in extensions. For example, on ARM, there is no equivalent to the `movemask` SSE/AVX x86 instruction.

The equations for the memory footprint provided in Sect. 3.2.2.6 also apply to our decoder implementation for embedded processors.

Comparison with Similar Works

Results were obtained using the ODROID-U3 board, which features a Samsung Exynos 4412 System on Chip (SoC) implementing an ARM Cortex A9 clocked at 1.7 GHz. Like in the previous sections, the decoders were only allowed to use one core. Table 3.5 shows the results for the proposed unrolled decoders and provides a comparison with [31]. As with their desktop CPU implementation of [32], inter-frame parallelism is used in the latter.

It can be seen that the proposed implementations provide better latency and greater throughput at native frequencies. Since the ARM CPU in the Samsung Exynos 4412 is clocked at 1.7 GHz while that in the NVIDIA Tegra 3 used in [31] is clocked at 1.4 GHz, we also provide linearly scaled throughput and latency numbers for the latter work, indicated by an asterisk (*) in the table. Compared to the scaled results of [31], the proposed decoder has 1.4–2.25 times the throughput and its latency is 25–36 times lower. The memory footprint of our proposed decoder is approximately 12 times lower than that of [31]. Both implementations are using 8-bit fixed-point values.

Table 3.5 Decoding polar codes with 8-bit fixed-point numbers on an ARM Cortex A9 using NEON

Code (N,k)	Decoder	Mem. Footprint (kBytes)	T/P (Mbps)		Latency (μs)
			Coded	Info	
(1024, 512)	[31]	38	70.5	35.3	232
	[31]*	38	80.6	42.9	191
	This work	3	113.1	56.6	9
(32 768, 29 492)	[31]	1216	33.1	29.8	15 844
	[31]*	1216	40.2	36.2	13 048
	This work	98	90.8	81.7	361

*Results linearly scaled for the clock frequency difference

3.4   Implementation on Graphical Processing Units

Most recent GPUs have the capability to do calculations that are not related to graphics. These GPUs are often called GPGPU. In this section, we describe our approach to implement software polar decoders in CUDA C [47] and present results for these decoders running on a NVIDIA Tesla K20c.

Most of the optimization strategies cited above could be applied or adapted to the GPU. However, there are noteworthy differences. Note that, when latency is mentioned below we refer to the decoding latency including the delay required to copy the data in and out of the GPU.

3.4.1   Overview of the GPU Architecture and Terminology

A NVIDIA GPU has multiple microprocessors with 32 cores each. Cores within the same microprocessor may communicate and share a local memory. However, synchronized communication between cores located in different microprocessors often has to go through the CPU and is thus costly and discouraged [15].

GPUs expose a different parallel programming model than general-purpose processors. Instead of SIMD, the GPU model is Single-Instruction Multiple-Threads (SIMT). Each core is capable of running a thread. A computational kernel performing a specific task is instantiated as a block. Each block is mapped to a microprocessor and is assigned one thread or more.

As it will be shown in Sect. 3.4.3, the latency induced by transferring data in and out of a GPU is high. To minimize decoding latency and maximize throughput, a combination of intra- and inter-frame parallelism is used for the GPU contrary to the CPUs where only the former was applied. We implemented a kernel that decodes a single frame. Thus, a block corresponds to a frame and attributing e.g. ten blocks to a kernel translates into the decoding of ten frames in parallel.

3.4.2   Choosing an Appropriate Number of Threads per Block

As stated above, a block can only be executed on one microprocessor but can be assigned many threads. However, when more than 32 threads are assigned to a block, the threads starting at 33 are queued for execution. Queued threads are executed as soon as a core is free.

Figure 3.3 shows that increasing the number of threads assigned to a block is beneficial only until a certain point is reached. For the particular case of a $(1024, 922)$ code, associating more than 128 threads to a block negatively affects performance. This is not surprising as the average node width for that code is low at 52.

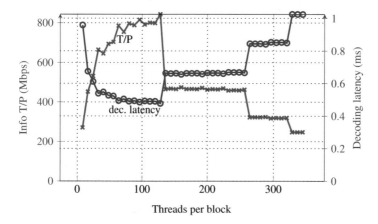

Fig. 3.3 Effect of the number of threads per block on the information throughput and decoding latency for a $(1024, 922)$ polar code where the number of blocks per kernel is 208

3.4.3   Choosing an Appropriate Number of Blocks per Kernel

Memory transfers from the host to the GPU device are of high throughput but initiating them induces a great latency. The same is also true for transfers in the other direction, from the device to the host. Thus, the number of distinct transfers have to be minimized. The easiest way to do so is to run a kernel on multiple blocks. For our application, it translates to decoding multiple frames in parallel as a kernel decodes one frame.

Yet, there is a limit to the number of resources that can be used to execute a kernel i.e. decode a frame. At some point, there will not be enough computing resources to do the work in one pass and many passes will be required. The NVIDIA Tesla K20c card features the Kepler GK110 GPU that has 13 microprocessors with 32 cores and 16 load and store units each [45]. In total, 416 arithmetic or logic operations and 208 load or store operations can occur simultaneously.

Figure 3.4 shows the latency to execute a kernel, to transfer memory from the host to the GPU and vice versa for a given number of blocks per kernel. The number of threads assigned per block is fixed to 128 and the decoder is built for a $(2048, 1707)$ polar code. It can be seen that the latency of memory transfers grows linearly with the number of blocks per kernel. The kernel latency however has local minimums at multiples of 208. We conclude that the minimal decoding latency, the sum of all three latencies illustrated in Fig. 3.4, is bounded by the number of load and store units.

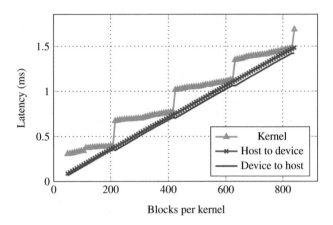

Fig. 3.4 Effect of the number of blocks per kernel on the data transfer and kernel execution latencies for a (2048, 1707) polar code where the number of threads per block is 128

3.4.4   On the Constituent Codes Implemented

Not all the constituent codes supported by the general purpose processors are beneficial to a GPU implementation. In a SIMT model, reduction operations are costly. Moreover, if a conditional execution leads to unbalanced threads, performance suffers. Consequently, all nodes based on the SPC codes, that features both characteristics, are not used in the GPU implementation.

Experiments have shown that implementing the SPC node results in a throughput reduction by a factor of 2 or more.

3.4.5   Shared Memory and Memory Coalescing

Each microprocessor contains shared memory that can be used by all threads in the same block. The NVIDIA Tesla K20c has 48 kB of shared memory per block. Individual reads and writes to the shared memory are much faster than accessing the global memory. Thus, intuitively, when conducting the calculations within a kernel, it seems preferable to use the shared memory as much as possible in place of the global memory.

However, as shown by Fig. 3.5, it is not always the case. When the number of blocks per kernel is small, using the shared memory provides a significant speedup. In fact, with 64 blocks per kernel, using shared memory results in a decoder that has more than twice the throughput compared to a kernel that only uses the global memory. Past a certain value of blocks per kernel though, solely using the global memory is clearly advantageous for our application.

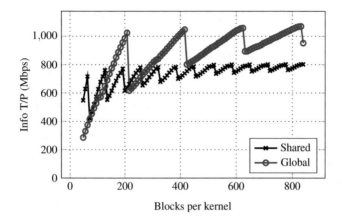

Fig. 3.5 Information throughput comparison for a $(1024, 922)$ polar code where intermediate results are stored in shared or global memory. The number of threads per block is 128

These results suggest that the GPU is able to efficiently schedule memory transfers when the number of blocks per kernel is sufficiently high.

3.4.6 Asynchronous Memory Transfers and Multiple Streams

Transferring memory from the host to the device and vice versa induces a latency that can be equal to the execution of a kernel. Fortunately, that latency can be first reduced by allocating pinned or page-locked host memory. As page-locked memory can be mapped into the address space of the device, the need for a staging memory is eliminated [47].

More significantly, NVIDIA GPUs with compute capability of 2.0 or above are able to transfer memory in and out of the device asynchronously. By creating three streams—sequences of operations that get executed in issue-order on the GPU—memory transfers and execution of the kernel can be overlapped, effectively multiplying throughput by a factor of 3.

This also increases the memory footprint by a factor of three. On the GPU, the memory footprint is

$$M_{\text{total (kbytes)}} = \frac{N(W_\beta + W_\alpha)BS}{8000}, \tag{3.6}$$

where B is the number of blocks per kernel—i.e. the number of frames being decoded simultaneously—S is the number of streams, and where W_β and W_α are the number of bits required to store a β and an α value, respectively. For best performance, as detailed in the next section, both β and α values are represented with floating-point values and thus $W_\beta = W_\alpha = 32$.

3.4.7 On the Use of Fixed-Point Numbers on a GPU

It is tempting to move calculations to 8-bit fixed-point numbers in order to speedup performance, just like we did with the other processors. However, GPUs are not optimized for calculations with integers. Current GPUs only support 32-bit integers. Even so, the maximum number of operations per clock cycle per multiprocessor as documented by NVIDIA [47] clearly shows that integers are third class citizens behind single- and double-precision floating-point numbers. As an example, Table 2 of [47] shows that GPUs with compute capability 3.5—like the Tesla K20c—can execute twice as many double-precision floating-point multiplications in a given time than it can with 32-bit integers. The same GPU can carry on six times more floating-point precision multiplications than its 32-bit integer counterpart.

3.4.8 Results

Table 3.6 shows the estimated information throughput and measured latency obtained by decoding various polar codes on a GPU. The throughput is estimated by assuming that the total memory transfer latencies are twice the latency of the decoding. This has been verified to be a reasonable assumption, using NVIDIA's profiler tool, when the number of blocks maximizes throughput.

Performing linear regression on the results of Table 3.6 indicates that the latency scales linearly with the number of blocks, leading to standard error values of 0.04, 0.04 and 0.14 for the $(1024, 922)$, $(2048, 1707)$ and $(4096, 3686)$ polar codes, respectively. In our decoder, a block corresponds to the decoding a single frame. The frames are independent of each other, and so are blocks. Thus, our decoder scales well with the number of available cores.

Table 3.6 Decoding polar codes on an NVIDIA Tesla K20c

(N, k)	Nbr of blocks	Info T/P (Mbps)	Latency (ms)
$(1024, 922)$	208	1022	0.6
	416	1046	1.1
	624	1060	1.6
	832	1070	2.2
$(2048, 1707)$	208	915	1.1
	416	936	2.2
	624	953	3.3
	832	964	4.5
$(4096, 3686)$	208	959	2.6
	416	1002	4.9
	624	1026	6.9
	832	1043	9.4

Furthermore, looking at Table 3.6 it can be seen that the information throughput is in the vicinity of a gigabit per second. Experiments have shown that the execution of two kernels can slightly overlap, making our throughput results of Table 3.6 worst-case estimations. For example, while the information throughput to decode 832 frames of a $(4096, 3686)$ polar code is estimated at 1043 Mbps in Table 3.6, the measured average value in NVIDIA's profiler tool was 1228 Mbps, a 18% improvement over the estimated throughput.

Our experiments have also shown that our decoders are bound by the data transfer speed that this test system is capable of. The PCIe 2.0 standard [1] specifies a peak data throughput of 64 Gbps when 16 lanes are used and once 8b10b encoding is accounted for. Decoding 832 frames of a polar code of length $N = 4096$ requires the transfer of 3 407 872 LLRs expressed as 32-bit floating-point numbers for a total of approximately 109 Mbits. Without doing any computation on the GPU, our benchmarks measured an average PCIe throughput of 45 Gbps to transfer blocks of data of that size from the host to the device and back. Running multiple streams and performing calculations on the GPU caused the PCIe throughput to drop to 40 Gbps. This corresponds to 1.25 Gbps when 32-bit floats are used to represent LLR inputs and estimated-bit outputs of the decoder. In light of these results, we conjecture that the coded throughput will remain approximately the same for any polar code as the PCIe link is saturated and data transfer is the bottleneck.

3.5 Energy Consumption Comparison

In this section the energy consumption is compared for all three processor types: the desktop processor, the embedded processor and the GPU. Unfortunately the Samsung Exynos 4412 SoC does not feature sensors allowing for power usage measurements of the ARM processor cores. The energy consumption of the ARM processor was estimated from board-level measurements. An Agilent E3631A DC power supply was used to provide the 5V input to the ODROID-U3 board and the current as reported by the power supply was used to calculated the power usage when the processor was idle and under load.

On recent Intel processors, power usage can be calculated by accessing the Running Average Power Limit (RAPL) counters. The LIKWID tool suite [64] is used to measure the power usage of the processor. Numbers are for the whole processor including the Dynamic Random-Access Memory (DRAM) package. Recent NVIDIA GPUs also feature on-chip sensors enabling power usage measurement. Steady state values are read in real-time using the NVIDIA Management Library (NVML) [46].

Table 3.7 compares the energy per information bit required to decode the $(2048, 1707)$ polar code. The SIMD-int8 implementation of our unrolled decoder is compared with that of the implementation in [32]. The former uses an Intel Core i7-4770S clocked at 3.1 GHz. The latter uses an Intel Core i7-4960HQ clocked

Table 3.7 Comparison of the power consumption and energy per information bit for the (2048, 1707) polar code

Decoder	Target	Mem. Footprint (kbytes)	Info. T/P (Gbps)	Latency (μs)	Power (W)	Energy (nJ/info. bit)
[32]	Intel Core i7-4960HQ[a]	144	2.2	26	13	6
This work	Intel Core i7-4770S	6	1.0	2	3	3
	Intel Core i7-4770S[a]	6	1.3	1	5	4
	ARM Cortex A9	6	0.1	14	0.8	7
	NVIDIA Tesla K20c	3408[b]	0.9	1100	108	118

[a]Results with Turbo enabled
[b]Amount required per stream. Three streams are required to sustain this throughput

at 3.6 GHz with Turbo enabled. The results for the ARM Cortex A9 embedded processor and NVIDIA Tesla K20c GPU are also included for comparison. Note that the GPU represents LLRs with floating-point numbers.

The energy per information bit is calculated with

$$E \ (J/\text{info. bit}) = \frac{P \ (W)}{\text{info. T/P} \ (bits/s)} .$$

It can be seen that the proposed decoder is slightly more energy efficient on a desktop processor compared to that of [32]. For that polar code, the latter offers twice the throughput but at the cost of a latency that is at least 13 times greater. However, the latter is twice as fast for that polar code. Decoding on the embedded processor offers very similar energy efficiency compared to the Intel processor although the data throughput is an order of magnitude slower. However, decoding on a GPU is significantly less energy efficient than any of the decoders running on a desktop processor.

The power consumption on the embedded platform was measured to be fairly stable with only a 0.1 W difference between the decoding of polar codes of lengths 1024 or 32 768.

3.6   Further Discussion

3.6.1   On the Relevance of the Instruction-Based Decoders

Some applications require excellent error-correction performance that necessitates the use of polar codes much longer than $N = 32\,768$. For example, Quantum Key Distribution benefits from frames of 2^{21} to 2^{24} bits [27]. At such lengths, current compilers fail to compile an unrolled decoder. However, the instruction-based decoders are very suitable and are capable of throughput greater than 100 Mbps with a code of length 1 million.

3.6.2 On the Relevance of Software Decoders in Comparison to Hardware Decoders

The software decoders we have presented are good for systems that require moderate throughput without incurring the cost of dedicated hardware solutions. For example, in a SDR communication chain based on USRP radios and the GNU Radio software framework, a forward error-correction (FEC) solution using our proposed decoders only consumes 5% of the total execution time on the receiver. Thus, freeing FPGA resources to implement functions other than FEC, e.g. synchronization and demodulation.

By building such a setup to demonstrate one of our software polar encoder and decoder pair, we were awarded the first place of the microsystems experimental demonstration competition at the 2015 edition of the Innovation Day event jointly organized by the IEEE Circuits and Systems Society and the Regroupement Stratégique en Microélectronique du Québec.

3.6.3 Comparison with LDPC Codes

LDPC codes are in widespread use in wireless communication systems. In this section, the error-correction performance of moderate-length polar codes is compared against that of standard LDPC codes [3]. Similarly, the performance of the state-of-the-art software LDPC decoders is compared against that of our proposed unrolled decoders for polar codes.

The fastest software LDPC decoders in literature are those of [23], which implements decoders for the 802.11n standard and present results for the Intel Core i7-2600 x86 processor. That wireless communication standard defines three code lengths: 1944, 1296, 648; and four code rates: 1/2, 2/3, 3/4, 5/6. In [23], LDPC decoders are implemented for all four codes rates with a code length of 1944. A layered offset-min-sum decoding algorithm with five iterations is used and early termination is not supported.

Figure 3.6 shows the FER of these codes using ten iterations of a flooding-schedule offset min-sum floating-point decoding algorithm which yields slightly better results than the five iteration layered algorithm used in [23]. The FER of polar codes with a slightly longer length of 2048 and matching code rates are also shown in Fig. 3.6.

Table 3.8 that provides the latency and information throughput for decoding 524 280 information bits using the state-of-the-art software LDPC decoders of [23] compared to our proposed polar decoders. To remain consistent with the result presented in [23], which used the Intel Core i7-2600 processor, the results in Table 3.8 use that processor as well.

While the polar code with rate $1/2$ offers a better coding gain than its LDPC counterpart, all other polar codes in Fig. 3.6 are shown to suffer a coding loss close

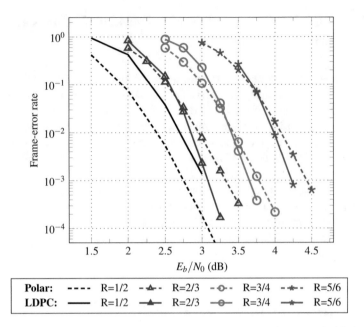

Fig. 3.6 Error-correction performance of the polar codes of length 2048 compared with the LDPC codes of length 1944 from the 802.11n standard

Table 3.8 Information throughput and latency of the polar decoders compared with the LDPC decoders of [23] when estimating 524 280 information bits on a Intel Core i7-2600

Decoder	N	Rate	Latency		Info. T/P (Mbps)
			total (ms)	per frame (μs)	
[23]	1944	1/2	17.4	N/A	30.1
		2/3	12.7	N/A	41.0
		3/4	11.2	N/A	46.6
		5/6	9.3	N/A	56.4
This work	2048	1/2	2.0	3.83	267.4
		2/3	1.0	2.69	507.4
		3/4	0.8	2.48	619.4
		5/6	0.6	2.03	840.9

to 0.25 dB at a FER of 10^{-3}. However, as Table 3.8 shows, there is approximately an order of magnitude advantage for the proposed unrolled polar decoders in terms of both latency and throughput compared to the LDPC decoders of [23].

3.7 Conclusion

In this chapter, we presented low-latency software polar decoders adapted to different processor architectures. The decoding algorithm is adapted to exploit different SIMD instruction sets for the desktop and embedded processors (SSE, AVX and NEON) or to the SIMT model inherent to GPU processors. The optimization strategies go beyond parallelisation with SIMD or SIMT. Most notably, we proposed to generate a branchless fully-unrolled decoder, to use compile-time specialization, and adopt a bottom-up approach by adapting the decoding algorithm and data representation to features offered by processor architectures. For desktop processors, we have shown that intra-frame parallelism can be exploited to get a very low-latency while achieving information throughputs greater than 1 Gbps using a single core. For embedded processors, the principle remains but the achievable information throughputs are more modest at 80 Mbps. On the GPU we showed that inter-frame parallelism could be successfully used in addition to intra-frame parallelism to reach better speed, and the impact of two critical parameters on the performance of the decoders was explored. We showed that given the right set of parameters, GPU decoders are able to sustain an information throughput around 1 Gbps while simultaneously decoding hundreds of frames. Finally, we showed that the memory footprint of our proposed decoder is at least an order of magnitude lower than that our the state-of-the-art polar decoder while being slightly more energy efficient. These results indicate that the proposed software decoders make polar codes interesting candidates for SDR applications. In fact, we won an award at a experimental demonstration competition by using our software solution in a over-the-air radio-communication setup.

Chapter 4
Unrolled Hardware Architectures for Polar Decoders

Abstract In this chapter, we demonstrate that polar decoders can achieve extremely high throughput values and retain moderate complexity. We present a family of architectures for hardware polar decoders using a reduced-complexity successive-cancellation decoding algorithm that employ unrolling. The resulting fully-unrolled architectures are capable of achieving a coded throughput in excess 400 Gbps and 1 Tbps on an FPGA or an ASIC, respectively—two to three orders of magnitude greater than current state-of-the-art polar decoders—while maintaining a competitive energy efficiency of 6.9 pJ/bit on ASIC. Moreover, the proposed architectures are flexible in a way that makes it possible to explore the trade-off between area, throughput and energy efficiency. We present the associated results for a range of pipeline depths, and code lengths and rates. We also discuss how the throughput and complexity of decoders are effected when implemented for an I/O-bound system.

4.1 Introduction

Conventional polar decoders implement one or a few specialized computational units and reuse them multiple times during the decoding process [33, 42, 48, 51, 55, 72]. It was shown in [54] and in Chap. 3 that unrolling the decoding process can lead to significant speed improvements in software polar decoders.

The goal of this chapter is to show how unrolling and pipelining the decoder tree can lead to hardware architectures that can achieve throughput values greater than 1 Tbps on a 28 nm CMOS technology ASIC operating at 1 GHz—three orders of magnitude faster than the state of the art. On FPGAs, the fastest architectures can reach hundreds of Gbps. Moreover, we present a family of architectures that offers a flexible trade-off between throughput, area and energy efficiency.

We start this chapter with Sect. 4.2 to provide a brief review of state-of-the-art polar decoder architectures. Section 4.3 presents the proposed family of architectures, and the operations and processing nodes used. It also specifies how code shortening can be used with the proposed architectures. Section 4.4 discusses the implementation and presents both FPGA and ASIC results for various code lengths and rates. Results are compared against state-of-the-art polar decoder implementations. The coded throughput of our decoders is shown to be in excess of 400 Gbps for a (2048, 1024) polar code decoded on an FPGA and of 1 Tbps for

© Springer International Publishing AG 2017   55
P. Giard et al., *High-Speed Decoders for Polar Codes*,
DOI 10.1007/978-3-319-59782-9_4

a $(1024, 512)$ polar code decoded on an ASIC, respectively two and three orders of magnitude over the current state of the art. Some power estimations are also provided for both FPGA and ASIC. Finally, Sect. 4.5 concludes this chapter.

Preliminary results of this work were presented in a letter [19]. In this chapter, we generalize the architecture into a family of architectures offering a flexible trade-off between throughput, area and energy efficiency, give more details on the unrolled architecture and provide more results. We also significantly improve the $(1024, 512)$ fully-unrolled deeply-pipelined polar decoder implementation results on all metrics. Finally, ASIC results for the 28 nm FD-SOI technology from STMicroelectronics are provided as well as power estimations for both FPGA and ASIC.

4.2   State-of-the-Art Architectures with Implementations

Most hardware polar decoder architectures presented in the literature, [33, 42, 48, 51, 55, 57, 72], use the SC decoding algorithm or an SC-based algorithm. These decoders require little logic area (ASIC) or resource usage (FPGA). As an example, the fastest of these SC-based decoders, the Fast-SSC decoder of [55], utilizes a processor-like architecture where the different units are used one to many times over the decoding of a frame. With the algorithmic improvements reviewed in Sect. 1.6, the Fast-SSC decoder was shown to be capable of achieving a 1.2 Gbps coded throughput (1.1 Gbps information throughput) at 108 MHz on an FPGA for a polar code with a length $N = 2^{15}$.

Recently, two polar decoders capable of achieving a coded throughput greater than 1 Gbps with a short $(1024, 512)$ polar code were proposed. An iterative BP fully-parallel decoder achieving a throughput of 4.7 Gbps at 300 MHz on a 65 nm CMOS ASIC was proposed in [49]. More recently, a fully-combinational, SC-based decoder with input and output registers was proposed in [14]. That decoder reaches a throughput of 2.9 Gbps at 2.79 MHz on a 90 nm CMOS ASIC and of 1.2 Gbps at 596 kHz on a 40 nm CMOS Xilinx Virtex 6 FPGA.

While these results are a significant improvement, their throughput, less than 6 Gbps, is still under the projected minimal peak throughput for future 5G communication standards [29, 43, 52]. Therefore, in this paper we propose a family of architectures capable of achieving one to three orders of magnitude greater throughput than the current state-of-the-art polar decoders.

4.3   Architecture, Operations and Processing Nodes

Similar to some decoders presented in the previous section, in order to significantly increase decoding throughput, our family of architectures does not focus on logic reuse but fully unrolls and pipelines the required calculations. A fully-unrolled decoder is a decoder where each and every operation or node required in estimating a codeword is instantiated with dedicated hardware. As an example, if a decoder for a specific polar code requires two executions of an F operation with a length of 8, a fully-unrolled decoder for that code will feature two F modules with inputs of size 8 instead of reusing the same block twice.

The idea of fully unrolling a decoder has previously been applied to decoders for other families of error-correcting codes. Notably, in [59, 68], the authors propose a fully-unrolled deeply-pipelined decoder for an LDPC code. Polar codes are more suitable to unrolling as they do not feature a complex interleaver like LDPC codes.

In this section, we provide details on the proposed family of architectures and describe the operations and processing nodes used by our architectures.

4.3.1 Fully Unrolled (Basic Scheme)

Building upon the work done on software polar decoders described in Chap. 3, we propose fully-unrolled hardware decoder architectures built for a specific polar code using a subset of the low-complexity Fast-SSC algorithm.

In the fully-unrolled architecture, all the nodes of a decoder tree exist simultaneously. Figure 4.2 shows a fully-unrolled decoder for the (8, 4) polar code illustrated as a decoder tree in Fig. 4.1b. White blocks represent operations in the Fast-SSC algorithm and the subscripts of their labels correspond to their input length N_v. Rep denotes a Repetition node, and C stands for the *Combine* operation. Grayed rectangles are registers. The clock and enable signals for those blocks are omitted for clarity. As it will be shown in Sect. 4.3.3, even with the multi-cycle paths, the enable signals for that decoder may always remain asserted without affecting the correctness as long as the input to the decoder remains stable for three clock cycles. This constitutes our basic scheme. On FPGA, it takes advantage of the fact that registers are available right after LUTs in logic blocks, meaning that adding a register after each operation does not require any additional logic block.

The code rate and frozen bit locations both affect the structure of the decoder tree and, in turn, the number of operations performed in a Fast-SSC decoder. However, as it will be shown in Sect. 4.4.4, the growth in logic usage or area for unrolled decoders remains $\mathcal{O}(N \log N)$, where N is the code length.

Fig. 4.1 Decoder trees for an (8, 4) polar code decoded with the (**a**) SSC and (**b**) Fast-SSC algorithms

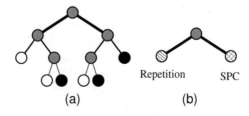

Repetition   SPC

(a)   (b)

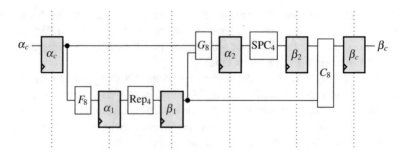

Fig. 4.2 Fully-unrolled decoder for a (8, 4) polar code. Clock and enable signals omitted for clarity

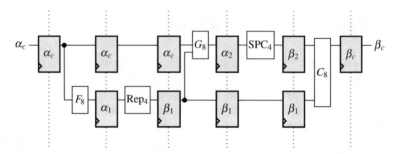

Fig. 4.3 Fully-unrolled deeply-pipelined decoder for a (8, 4) polar code. Clock signals omitted for clarity

4.3.2 Deeply Pipelined

In a deeply-pipelined architecture, a new frame is loaded into the decoder at every clock cycle. Therefore, a new estimated codeword is output at each clock cycle as each register is active at each rising edge of the clock (no enable signal required). In that architecture, at any point in time, there are as many frames being decoded as there are pipeline stages. This leads to a very-high throughput at the cost of high memory requirements. Some pipeline stage paths do not contain any processing logic, only memory. They are added to ensure that the different messages remain synchronized. These added memories yield register chains, or SRAM blocks, as will be shown in Sect. 4.3.5.

The unrolled decoder of Fig. 4.2 can be transformed into a deeply-pipelined decoder by adding four registers. Two registers are needed to retain the channel LLRs, denoted α_c in the figure, during the second and third clock cycles. Similarly, two registers have to be added for the persistence of the hard-decision vector β_1 over the fourth and fifth clock cycles. Making these modifications results in the fully-unrolled deeply-pipelined decoder shown in Fig. 4.3. Figure 4.4 shows another example of a fully-unrolled deeply-pipelined decoder, but for a (16, 14) polar code featuring more operations and node types compared to Fig. 4.3, where I denotes a Rate-1 node.

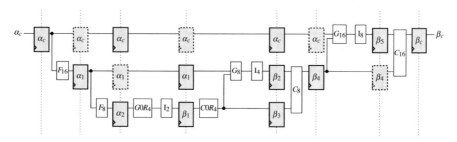

Fig. 4.4 Fully-unrolled deeply-pipelined decoder for a (16, 14) polar code. Clock signals omitted for clarity

For this architecture, the amount of memory required is quadratic in code length and, similarly to resource usage, affected by rate and frozen bit locations. As will be shown in Sect. 4.4, this growth in memory usage limits the proposed deeply-pipelined architecture to codes of moderate lengths, under 4096 bits, at least for implementations using the target FPGA.

Information throughput is defined as PfR bps, where P is the width of the output bus in bits, f is the execution frequency in Hz and R is the code rate. In a deeply-pipelined architecture, P is assumed to be equal to the code length N. The decoding latency depends on the frozen bit locations and the constrained maximum width for all processing nodes, but is less than $N \log_2 N$. In our experiments, with the operations and optimizations described below, the decoding latency never exceeded $N/2$ clock cycles.

4.3.3 Partially Pipelined

In a deeply-pipelined architecture, a significant amount of memory is required for data persistence. That memory quickly increases with the code length N. Instead of loading a new frame into the decoder and estimating a new codeword at every cycle, we propose a compromise where the unrolled decoder can be partially pipelined to reduce the required memory. Let $\mathscr{I}$ be the initiation interval, where a new estimated codeword is output every $\mathscr{I}$ clock cycles. The case where $\mathscr{I} = 1$ translates to a deeply-pipelined architecture.

Setting $\mathscr{I} > 1$ leads to a significant reduction in the memory requirements. An initiation interval of $\mathscr{I}$ translates to an effective required register chain length of $\lceil L/\mathscr{I} \rceil$ instead of L, where L is the length of the register chain. Using $\mathscr{I} = 2$ leads to a ~50% reduction in the amount of memory required for that section of the circuit. This reduction applies to all register chains present in the decoder.

The unrolled decoder of Fig. 4.2 can be seen as a partially-pipelined decoder with an initiation interval $\mathscr{I} = 3$. A partially-pipelined decoder with $\mathscr{I} = 2$ can be obtained for a (16, 14) polar code by removing the dotted registers in Fig. 4.4, leading to the decoder shown in Fig. 4.5.

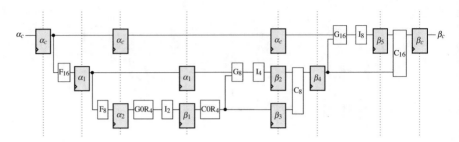

Fig. 4.5 Fully-unrolled partially-pipelined decoder for a (16, 14) polar code with $\mathscr{I} = 2$. Clock and enable signals omitted for clarity

The initiation interval $\mathscr{I}$ can be increased further in order to reduce the memory requirements, but only up to a certain limit (corresponding to the basic scheme). We call that limit the maximum initiation interval $\mathscr{I}_{\max}$, and its value depends on the decoder tree. By definition, the longest register chain in a fully-unrolled decoder is used to preserve the channel LLRs α_c. Hence, the maximum initiation interval corresponds to the number of clock cycles required for the decoder to reach the last operation in the decoder tree that requires α_c, G_N, the operation calculated when going down the right edge linking the root node to its right-hand-side child. Once that G_N operation is completed, α_c is no longer needed and can be overwritten. As an example, consider the (8, 4) polar decoder illustrated in Fig. 4.2. As soon as the switch to the right-hand side of the decoder tree occurs, i.e. when G_8 is traversed, the register containing the channel LLRs α_c can be updated with the LLRs for the new frame without affecting the remaining operations for the current frame. Thus the maximum initiation interval, $\mathscr{I}_{\max}$, for that decoder is 3.

The resulting information throughput is $PfR/\mathscr{I}$ bps, where $\mathscr{I}$ is the initiation interval. Note that this new definition can also be used for the deeply-pipelined architecture. The decoding latency remains unchanged compared to the deeply-pipelined architecture.

The partially-pipelined architecture requires a more elaborate controller than the deeply-pipelined architecture. For both fully- and partially-pipelined architectures, the controller generates a done signal to indicate that a new estimated codeword is available at the output. For the partially-pipelined architecture, the controller also contains a counter with maximum value of $(\mathscr{I} - 1)$ which generates the $\mathscr{I}$ enable signals for the registers. An enable signal is asserted only when the counter reaches its value, in $[0, \mathscr{I} - 1]$, otherwise it remains deasserted. Each register uses an enable signal corresponding to its location in the pipeline modulo $\mathscr{I}$. As an example, let us consider the decoder of Fig. 4.5, i.e. $\mathscr{I}$ is set to 2. In that example, two enable signals are created and a simple counter alternates between 0 and 1. The registers storing the channel LLRs α_c are enabled when the counter is equal to 0 because their input resides on the even (0, 2 and 4) stages of the pipeline. On the other hand, the two registers holding the α_1 LLRs are enabled when the counter is equal to 1 because their inputs are on odd (1 and 3) stages. The other registers follow the same rule.

The required memory resources could be further reduced by performing the decoding operations in a combinational manner, i.e. by removing all the registers except the ones labeled α_c and β_c, as in [14]. However, the resulting reachable frequency is too low for the desired throughput level.

4.3.4 Operations and Processing Nodes

In order to keep the critical paths as short as possible, only a subset of the operations and processing nodes proposed in the original Fast-SSC algorithm are used. Furthermore, for some nodes, the maximum processing node length N_v is constrained to smaller values than the ones used in [55].

Notably, the Repetition and SPC nodes are limited to $N_v = 8$ and 4, respectively. The remainder of the operations—F, G, $G0R$, *Combine*, $C0R$—are not constrained, as their lengths do not affect the critical paths. While it was not required in the original Fast-SSC algorithm, our architecture includes a Rate-1 processing node, implementing (1.4). That Rate-1 node is not constrained in length either.

In order to reduce latency and resource usage, improvements were made to some operations and nodes. They are detailed below.

$C0R$ Operations: A $C0R$ operation is a special case of a *Combine* operation (1.3) where the left-hand-side constituent code, β_l, is a Rate-0 node. Thus, the operation is equivalent to copying the estimated hard values from the right-hand-side constituent code over to the left-hand side.

  In other words, a $C0R$ does not require any logic, it only consists of wires. All occurrences of that operation were thus merged with the following modules, saving a clock cycle without negatively impacting the maximum clock frequency and reducing memory.

Rate-1 or Information Nodes: With a fixed-point number representation, a Rate-1 (or Information) node amounts to copying the most significant bit of the input LLRs. Similarly to the $C0R$ operation, the Information node does not require any logic and is equivalent to wires.

  Contrary to the $C0R$ operation though, we do not save a clock cycle by prepending the Information node to its consumer node. Instead, the register storing LLRs at the output of its producer is removed and the Information node is appended, along with its register used to store the hard decisions. Not only is the decoding latency reduced by a clock cycle, but a register storing LLR values is removed.

Repetition Nodes: The output of a Repetition node is a single bit estimate. The systematic polar decoder of [55] copies that estimated information bit N_v times to form the estimated bit vector, before storing it. In our implementation, we store only one bit that is later expanded just before a consumer requires it. This reduces the width of register chains carrying bit estimates generated by Repetition nodes, thus decreasing resource usage.

4.3.5   *Replacing Register Chains with SRAM Blocks*

As the code length N grows, long register chains start to appear in the decoder, especially with a smaller $\mathscr{I}$. In order to reduce the number of registers required, register chains can be converted into SRAM blocks.

Consider the register chain of length 6 used for the persistence of the channel LLRs α_c in the fully-unrolled deeply-pipelined $(16, 14)$ decoder shown in top row of Fig. 4.4. That register chain can be replaced by an SRAM block with a depth of 6 along with a controller to generate the appropriate read and write addresses. Similar to a circular buffer, if the addresses are generated to increase every clock cycle, the write address is set to be one position ahead of the read address.

SRAM blocks can replace register chains in a partially-pipelined architecture as well. In both architectures, the SRAM block depth has to be equal or greater than the register chain length. The same constraint applies to the width.

In scenarios where narrow SRAM blocks are not desirable, register chains can be merged to obtain a wider SRAM block even if the register chains do not have the same length. If the lengths of two register chains to be merged differ, the first registers in the longest chain are preserved, and only the remaining registers are merged with the other chain.

4.4   Implementation and Results

4.4.1   *Methodology*

In our experiments, decoders are built with sufficient memory to accommodate storing an extra frame at the input, and to preserve an estimated codeword at the output. As a result, the next frame can be loaded while a frame is being decoded. Similarly, an estimated codeword can be read while the next frame is being decoded. To facilitate comparison between the fully and partially-pipelined architectures, we define decoding latency to only include the time required for the decoder to decode a frame; loading channel LLRs and offloading estimated codewords are excluded from the calculations.

The quantization used was determined by running fixed-point simulations with bit-true models of the decoders. A smaller number of bits is used to store the channel LLRs compared to that of the other LLRs used in the decoder. All LLRs share the same number of fractional bits. We denote quantization as $Q_i.Q_c.Q_f$, where Q_c is the total number of bits to store a channel LLR, Q_i is total the number of bits used to store internal LLRs and Q_f is the number of fractional bits in both. Figure 4.6 shows the effect of quantization on the error-correction performance of a $(1024, 512)$ polar code modulated with BPSK and transmitted over an AWGN channel. Note that more fixed-point combinations are illustrated compared to previous chapters.

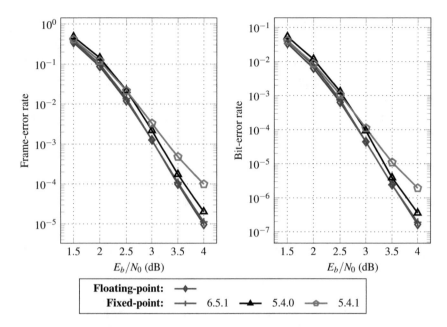

Fig. 4.6 Effect of quantization on the error-correction performance of a (1024, 512) polar code

Looking at Fig. 4.6, it can be seen that using $Q_i.Q_c.Q_f$ equal to 5.4.0 results in a 0.1 dB performance degradation at a BER of 10^{-6}. Thus we used that quantization for the hardware results.

FPGA results are for an Altera Stratix IV EP4SGX530-KH40C2 to facilitate comparison against most polar decoder implementations in the literature. That FPGA features 637 440 LUTs, 424 960 registers and 21 233 664 bits of SRAM. All FPGA results are worst-case using the slow 900 mV 85 °C timing model. Better results are to be expected if more recent FPGAs were to be targeted. ASIC synthesis results for the 28 nm FD-SOI CMOS technology from STMicroelectronics are obtained by running Synopsys Design Compiler in topographical mode for a typical library at 1.0 V and 125 °C, and a timing constraint set to 1 ns. Furthermore, on ASIC, only registers are used as we did not have access to an SRAM compiler.

4.4.2   Effect of the Initiation Interval

In this section, we explore the effect of the initiation interval on the implementation of the fully-unrolled architecture. The decoders are built for the same (1024, 512) polar code used in [19], although many improvements were made since the publication of that work (see Sect. 4.3.4). Regardless of the initiation interval, all decoders use 5.4.0 quantization and have a decoding latency of 364 clock cycles.

Table 4.1 Decoders for a $(1024, 512)$ polar code with various initiation interval $\mathcal{I}$ implemented on an FPGA

$\mathcal{I}$	LUTs	Regs.	RAM (kbits)	f (MHz)	T/P (Gbps)	Latency (μs)
1	136 874	188 071	83.92	248	254.1	1.47
2	124 532	166 361	9.97	238	121.7	1.53
3	114 173	152 182	4.68	208	71.1	1.75
4	110 381	145 000	0	203	52.1	1.79
50	86 998	65 618	0	218	4.5	1.67
167	93 225	43 236	0	239	1.5	1.52

Table 4.2 Decoders for a $(1024, 512)$ polar code with various initiation interval $\mathcal{I}$ implemented on an ASIC

$\mathcal{I}$	Tot. Area (mm^2)	Mem. Area (mm^2)	f (MHz)	T/P (Gbps)	Latency (μs)
1	4.627	3.911	1245	1274.9	0.29
2	3.326	3.150	1020	522.2	0.36
3	2.314	2.138	1005	343.0	0.36
4	1.665	1.063	1003	256.8	0.36
50	0.366	0.143	1003	20.5	0.36
167	0.289	0.089	1003	6.2	0.36

Tables 4.1 and 4.2 show the results for various initiation intervals on the FPGA and ASIC implementations, respectively. Besides the effect on coded throughput, increasing the initiation interval causes a significant reduction in the FPGA resources required or of the ASIC area. On FPGA, while the throughput is approximately cut in half, using $\mathcal{I} = 2$ reduces the number of required LUTs, registers and RAM bits by 9, 12 and 88%, respectively, compared to the deeply-pipelined decoder. Also on FPGA, with a throughput over 50 Gbps, using an initiation interval as small as four removes the need for any SRAM blocks, while the usage of LUTs and registers decreases by 20 and 23%, respectively. Finally, from Table 4.1, if a coded throughput of 1.5 Gbps is sufficient for the application, $\mathcal{I} = 167$ will result in savings of 32, 77 and 100% in terms of LUTs, registers and RAM bits, compared to the deeply-pipelined architecture ($\mathcal{I} = 1$). On ASIC, the area is largely dominated by registers and thus increasing the initiation interval has great effect on the total area as shown in Table 4.2. For example, using $\mathcal{I} = 50$ results in an area that is more than 12 times smaller, at the cost of a throughput that is 50 times lower.

As expected, increasing the initiation interval $\mathcal{I}$ offers a diminishing return as it gets closer to the maximum of 167. Table 4.1 also shows that on FPGA increasing $\mathcal{I}$ first reduces the maximum execution frequency but, eventually, it reincreases almost back to the value it had with $\mathcal{I} = 1$. Inspection of the critical paths reveals that this frequency increase is a result of shorter wire delays. As the number of LUTs and registers decreases with an increasing $\mathscr{I}$, at some point, it becomes easier to use resources that are close to each other.

4.4.3 Comparison with State-of-the-Art Decoders

In this section, we compare our work with that of the fastest state-of-art polar decoder implementations: [14, 17, 49]. The work of [17] was presented in Chap. 2. In [49], ASIC results are provided. The work of [17] and [14] provide results for both ASIC and FPGA implementations. The BP decoder of [49] is an iterative decoder utilizing early termination to improve the average throughput. However, as it does not include the necessary buffers to accommodate that functionality, we add the sustainable throughput for consistency.

Table 4.3 shows that regardless of the implementation technology, our family of architectures can deliver from one to three orders of magnitude greater coded throughput. On ASIC, the latency is about the same for all implementations.

Normalizing the results of [14, 17, 49] to 28 nm CMOS technology with a supply voltage of 1.0 V, the coded throughput is still one or two orders of magnitude greater. Latency is approximately $3/2$ to 2 times that of [49] and three to four times greater than [14]. Looking at the energy efficiency for the deeply-pipelined architecture on ASIC, the proposed decoder is six times more efficient than [49], 3.5 times more efficient than [17] and about two times less efficient than [14]. In terms of area efficiency, the same decoder is 4.2, 16 and 9.4 times more efficient than that of [17, 49] and [14], respectively.

Table 4.4 compares our proposed fully-unrolled partially-pipelined architecture, with the maximum initiation interval $\mathscr{I}_{max} = 167$, against the fastest FPGA implementations of [14, 17]. The work of [14] is marked with (†) as these results are for a different FPGA, the Xilinx Virtex-6 XC6VLX550T. Note however that this Xilinx FPGA is implemented in 40 nm CMOS technology and features 6-input LUTs, like the Altera Stratix IV FPGA.

It can be seen that a decoder built with one of our proposed architectures can achieve nearly three times the throughput of both [17] and [14] with a slightly lower latency. In terms of resources, compared to the decoder of [17], our decoder requires almost four and seven times the number of LUTs and registers, respectively. Note however that we do not require any RAM for the proposed implementation, while the other decoder from Chap. 2 uses 44 kbits. Compared to the SC decoder of [14], our decoder requires less than half the LUTs, but needs more than seven times the number of registers. It should be noted that the decoder of [14] does not contain the necessary memory to load the next frame while a frame is being decoded, nor the necessary memory to offload the previously estimated codeword as decoding is taking place.

Table 4.3 Comparison with state-of-the-art polar decoders

	This work						[17]	[49]ᵃ	[14]
Algorithm	Fast-SSC						Fast-SSC	BP	SC
Code	(1024, 512)						(1024, 512)	(1024, 512)	(1024, k)
IC Type	FPGA			ASIC			ASIC	ASIC	ASIC
Technology	40 nm			28 nm			65 nm	65 nm	90 nm
Init. interval ($\mathcal{I}$)	167	50	1	167	50	1	–	–	–
Supply (V)	0.9	0.9	0.9	1.0	1.0	1.0	1.0	1.0	1.3
Area (mm²)	–	–	–	0.29	0.37	4.63	0.69	1.48	3.21
Frequency (MHz)	239	218	248	1003	1003	1245	600	300	2.5
Latency (μs)	1.5	1.7	1.5	0.4	0.4	0.3	0.4	0.5	0.4
T/P (Gbps)	1.5	4.5	254	6.2	20.5	1275	3.7	4.7 @ 4 dB	2.9
Sust. T/P (Gbps)	1.5	4.5	254	6.2	20.5	1275	3.7	2.0	2.9
Area Eff. (Gbps/mm²)	–	–	–	21.45	56.01	275.53	5.4	3.18 @ 4 dB	0.8
Power (mW)	1420	1454	5532	169	271	8793	215	478	191
Energy (pJ/bit)	946.7	323.2	21.8	27.3	13.2	6.9	57.5	102.1	74.5
Normalized results for 28 nm and 1.0 V									
Area (mm²)	–	–	–	0.29	0.37	4.63	0.13	0.27	0.31
Frequency (MHz)	341	311	354	1003	1003	1245	1392	696	8.0
Latency (μs)	1.1	1.2	1.1	0.4	0.4	0.3	0.2	0.2	0.1
Sust. T/P (Gbps)	2.1	6.4	363	6.2	20.5	1275	8.6	4.6	9.2
Area Eff. (Gbps/mm²)	–	–	–	21.45	56.01	275.53	66.2	17.2	29.6
Power (mW)	1227	1257	4781	169	271	8793	93	206	35
Energy (pJ/bit)	818.1	279.2	18.8	27.3	13.2	6.9	25.0	44.8	3.8

ᵃMeasurement results

Table 4.4 Comparison with other FPGA implementations

Impl.	LUTs	Regs.	RAM (kbits)	f (MHz)	T/P (Gbps)	Latency (μs)
This work	93 225	43 236	0	239	1.5	1.5
[17]	23 353	5814	44	103	0.6	1.6
[14]†	193 456	6151	N/A	1	0.6	3.4

Table 4.5 Deeply-pipelined decoders for polar codes of various lengths with rate $R = {}^1\!/_2$ implemented on an FPGA

N	LUTs	Regs.	RAM (kbits)	f (MHz)	T/P (Gbps)	Latency (μs)
128	9917	18 543	0	357	45.7	0.21
256	27 734	44 010	0	324	83.0	0.41
512	64 723	105 687	4	275	141.1	0.74
1024	136 874	188 071	84	248	254.1	1.47
2048	217 175	261 112	5362	203	415.7	3.21

Table 4.6 Deeply-pipelined decoders for polar codes of various lengths with rate $R = {}^1\!/_2$ implemented on an ASIC

N	Tot. Area (mm^2)	Log. Area (mm^2)	Mem. Area (mm^2)	f (MHz)	T/P (Gbps)	Latency (ns)
128	0.125	0.027	0.098	1383	177.0	55.0
256	0.412	0.079	0.332	1353	346.4	99.0
512	1.263	0.217	1.045	1328	679.9	156.6
1024	4.627	0.715	3.911	1245	1274.9	292.4

4.4.4   Effect of the Code Length and Rate

Results for other polar codes are presented in this section where we show the effect of the code length and rate on performance and resource usage.

Tables 4.5, 4.6, 4.7 and 4.8 show the effect of the code length on resource usage, coded throughput, and decoding latency for polar codes of short to moderate lengths. Tables 4.5 and 4.6 contain results for the fully-unrolled deeply-pipelined architecture ($\mathscr{I} = 1$) and the code rate R is fixed to $^1\!/_2$ for all polar codes. Tables 4.7 and 4.8 contain results for the fully-unrolled partially-pipelined architecture where the maximum initiation interval ($\mathscr{I}_{\max}$) is used and the code rate R is fixed to $^5\!/_6$.

As shown in Tables 4.5 and 4.6, with a deeply-pipelined architecture, both the logic usage and memory requirements are close to being quadratic in code length N.

On FPGAs, the decoders for the three longest codes of Table 4.5 are capable of a coded throughput greater than 100 Gbps. Notably, the $N = 2048$ code reaches 400 Gbps. On ASIC, a throughput exceeding 1 Tbps can be achieved with a decoder for a polar code of length $N = 1024$ as shown in Table 4.6. The decoder for the

Table 4.7 Partially-pipelined decoders with initiation interval set to $\mathscr{I}_{max}$ for polar codes of various lengths with rate $R = 5/6$ implemented on an FPGA

N	$\mathscr{I}$	LUTs	Regs.	f (MHz)	T/P (Gbps)	Latency (μs)
1024	206	75 895	42 026	236	1.17	1.11
2048	338	165 329	75 678	220	1.33	2.02
4096	665	364 320	172 909	123	0.76	7.04

Table 4.8 Partially-pipelined decoders with initiation interval set to $\mathscr{I}_{max}$ for polar codes of various lengths with rate $R = 5/6$ implemented on an ASIC clocked at 1 GHz

N	$\mathscr{I}$	Tot. Area (mm^2)	Log. Area (mm^2)	Mem. Area (mm^2)	T/P (Gbps)	Latency (μs)
1024	206	0.230	0.169	0.070	5.0	0.26
2048	338	0.509	0.345	0.164	6.1	0.44
4096	665	1.192	0.820	0.372	6.2	0.87

$(2048, 1024)$ polar code could not be synthesized for ASIC on our server due to insufficient memory.

Table 4.7 shows that for a partially-pipelined decoder where the initiation interval is set to $\mathscr{I}_{max}$, it is possible to fit a code of length $N = 4096$ on the Stratix IV GX 530. The amount of RAM required is not illustrated in the table as none of the decoders are using any of the available RAM. Also note that no LUTs are used as memory. In other words, for pipelined decoders using $\mathscr{I}_{max}$ as the initiation interval, registers are the only memory resources needed. Table 4.7 also shows that these maximum initiation intervals lead to a much more modest throughput. In the case of the $(4096, 3413)$ polar code, we can see a major latency increase compared to the shorter codes. This latency increase can be explained by the maximum clock frequency drop which in turn can be explained by the fact that 94% of the total available logic resources in that FPGA were required to implement this decoder.

At some point on FPGAs, a fully-unrolled architecture is no longer advantageous over a more compact architecture like the one of [55]. With the Stratix IV GX 530 as an FPGA target, a fully-unrolled decoder for a polar code of length $N = 4096$ is too complex to provide good throughput and latency. Even with the maximum initiation interval, 94% of the logic resources are required for a coded throughput under 1 Gbps. By comparison, a decoder built with the architecture of [55] would result in a coded throughput in the vicinity of 1 Gbps at 110 MHz. Targeting a more recent FPGA could lead to different results and conclusions.

On ASIC, both the memory and total area scale linearly with N for a partially-pipelined architecture with $\mathscr{I}_{max}$. The results of Table 4.8 also show that it was possible to synthesize ASIC decoders for larger code lengths than what was possible with a deeply-pipelined architecture.

The effect of using different code rates for a polar code of length $N = 1024$ is shown in Tables 4.9 and 4.10. We note that the higher-rate codes do not have

Table 4.9 Deeply-pipelined decoders for polar codes of length $N = 1024$ with common rates implemented on an FPGA

R	LUTs	Regs.	RAM (bits)	f (MHz)	T/P (Gbps)	Latency (CCs)	(μs)
1/2	136 874	188 071	83 924	248	254.1	364	1.47
2/3	137 230	183 957	73 020	250	256.0	326	1.30
3/4	151 282	204 479	83 288	227	232.7	373	1.64
5/6	145 659	198 876	82 584	229	234.4	323	1.41

Table 4.10 Deeply-pipelined decoders for polar codes of length $N = 1024$ with common rates implemented on an ASIC

R	Tot. Area (mm^2)	Log. Area (mm^2)	Mem. Area (mm^2)	f (MHz)	T/P (Gbps)	Latency (CCs)	(ns)
1/2	4.627	0.715	3.911	1245	1274.9	364	292.4
2/3	4.896	0.740	4.156	1300	1331.2	326	250.8
3/4	5.895	0.872	5.023	1245	1274.9	373	299.6
5/6	5.511	0.816	4.694	1361	1393.7	323	237.3

noticeably lower latency compared to the rate-$1/2$ code, contrary to what was observed in [55]. This is due to limiting the width of SPC nodes to 4 in this work, whereas it was left unbounded in [17, 20, 55]. The result is that long SPC codes are implemented as trees whose left-most child is a width-4 SPC node and the others are all rate-1 nodes. Thus, for each additional stage ($\log_2 N_v - \log_2 N_{SPC}$) of an SPC code of length $N_v > N_{SPC}$, four nodes with a total latency of 3 CCs are required: F, G followed by I, and *Combine*. This brings the total latency of decoding a long SPC code to $3(\log_2 N_v - \log_2 N_{SPC}) + 1$ CCs compared to $\lceil N_v/\mathscr{P} \rceil + 4$ in [55], where $\mathscr{P}$ is the number of LLRs that can be read simultaneously (256 was a typical value for $\mathscr{P}$ in [55]).

Figure 4.7 gives a graphical overview of the maximum resource usage requirements on FPGA for a given achievable coded throughput. The fully-unrolled deeply- and partially-pipelined decoders were taken from Tables 4.1 and 4.5, respectively. The resynthesized polar decoder of [55] is also included for reference. The red asterisks show that with a deeply-pipelined decoder architecture (initiation interval $\mathscr{I} = 1$), the coded throughput increases at a higher rate than the maximum resource usage as the code length N increases. The blue diamonds illustrate the effect of various initiation intervals for the same (1024, 512) polar code. We see that decreasing $\mathscr{I}$ leads to increasingly interesting implementation alternatives, as the gains in throughput are obtained at the expense of a smaller increase in the maximum resource usage. When it can be afforded and that the FPGA input data rate is sufficient, the extra 2.9% in maximum resource usage allows doubling the throughput, from $\mathscr{I} = 2$ to $\mathscr{I} = 1$.

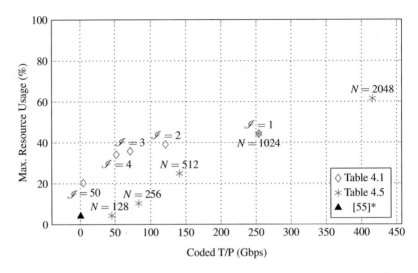

Fig. 4.7 Overview of the maximum FPGA resource usage and coded throughput for some partially-pipelined (Table 4.1) and deeply-pipelined (Table 4.5) polar decoders. The resynthesized polar decoder of [55] is also included for reference

4.4.5   On the Use of Code Shortening in an Unrolled Decoder

Code shortening is a well-known technique used to create a rate- and length-flexible error-correction system. Multiple such schemes were proposed for use with polar codes [38, 41, 67]. The technique presented in [38] could be used with this work. It starts from a systematic (N, k) polar code and shortens it by h bits, resulting in a (N_s, k_s) polar code, where $N_s = N - h$ and $k_s = k - h$. At the encoder, the h information bit locations of highest indices are set to a predetermined value (usually 0). After encoding, these bits are discarded from x before being transmitted over the channel. At the decoder, the corresponding soft-inputs channel values are set to a certain 0 or 1, depending on the predetermined value used at the encoder.

To support a shortened $(N - h, k - h)$ polar code, an unrolled decoder requires additional circuitry at the input to insert LLR values for the discarded bits—the maximum LLR value when the discarded bits are assumed to be 0. Since, the scheme of [38] chooses the information bits with largest indices to discard, the routing overhead would be minor.

4.4.6   I/O Bounded Decoding

The family of architectures that we propose requires tremendous throughput at the input of the decoder, especially with a deeply-pipelined architecture. For example, if a quantization of $Q_c = 4$ bits is used for channel LLRs, for every estimated bit, four times as many bits have to be loaded into the decoder. In other words, the total data rate is five times that of the output. This can be a significant challenge on both FPGAs and ASICs.

On FPGA, if 38 of the 48 high-speed transceivers (approximately $4/5$) featured on a Stratix IV GX are to be used to load the channel LLRs and the remainder to output the estimated codewords, the maximum theoretical input data rate achievable will be of 323 Gbps. On the more recent Stratix V GX, using 53 of the 66 transceivers at their peak data rate of 14.1 Gbps sums up to 747 Gbps available for input. However, the fully-unrolled deeply-pipelined (1024, 512) and (2048, 1024) polar decoders discussed above require an input data rate that is over 1 Tbps.

If only for that reason, partially-pipelined architectures are certainly more attractive, at least using current FPGA technology. Notice however that data rates in the vicinity of 1 Tbps are expected to be reachable in the incoming Xilinx UltraScale [70] and Altera Generation 10 [5] families of FPGAs. On ASICs, the number of high-speed transceivers is not fixed and a custom solution can be built.

4.5 Conclusion

In this chapter we presented a new family of architectures for fully-unrolled polar decoders. With an initiation interval that can be adjusted, these architectures make it possible to find a trade-off between area (or resource usage) and achievable throughput without affecting decoding latency. We showed that a fully-unrolled deeply-pipelined decoder implemented on an FPGA can achieve a throughput greater than 400 Gbps, which is two orders of magnitude greater than state-of-the-art polar decoders while maintaining a good latency. On ASICs, we showed that the proposed fully-unrolled deeply-pipelined decoders could achieve a throughput that would be two or three orders of magnitude greater than the state-of-the-art decoders with an order of magnitude better normalized area efficiency and a competitive energy efficiency. One of the proposed decoder has a coded throughput in excess of 1 Tbps at 6.9 pJ/bit on ASICs. We believe that these architectures make polar codes a promising candidate for future 5G communications.

Chapter 5
Multi-Mode Unrolled Polar Decoding

Abstract Unrolled decoders are architectures that provide the greatest decoding speed, by orders of magnitude compared to their more compact counterparts. However, unrolled decoders are built for a specific, fixed, code. In this chapter, we present a new method to enable the use of multiple code lengths and rates in a fully-unrolled polar decoder architecture. This novel method leads to a length- and rate-flexible decoder while retaining the very high speed typical to those decoders. We present results for two versions of a multi-mode decoder supporting eight and ten different polar codes, respectively. Both are capable of a peak throughput of 25.6 Gbps. For each decoder, the energy efficiency for the longest supported polar code is shown to be of 14.8 pJ/bit at 250 MHz and of 8.8 pJ/bit at 500 MHz on an ASIC built in 65 nm CMOS technology.

5.1 Introduction

In the previous chapter (and in [21]), unrolled hardware architectures for polar decoders were proposed. Results showed a very high throughput, greater than 1 Tbps. However, these architectures are built for a fixed polar code i.e. the code length or rate cannot be modified at execution time. This is a major drawback for most modern wireless communication applications that largely benefit from the support of multiple code lengths and rates.

The goal of this chapter is to show how an unrolled decoder built specifically for a polar code, of fixed length and rate, can be transformed into a multi-mode decoder supporting many codes of various lengths and rates. More specifically, we show how decoders for moderate-length polar codes contain decoders for many other shorter—but practical—polar codes of both high and low rates. The required hardware modifications are detailed, and ASIC synthesis and power estimations are provided for the 65 nm CMOS technology from TSMC. Results show a peak information throughput greater than 25 Gbps at 250 MHz in 4.29 mm^2 or at 500 MHz in 1.71 mm^2. Latency is of 2 μs and 650 ns for the former and latter.

The remainder of this chapter starts with Sects. 5.2 and 5.3 where a supporting decoder tree example is provided along with its unrolled hardware implementation. Section 5.4 then explains the concept, hardware modifications and other practical considerations related to the proposed multi-mode decoder. Error-correction perfor-

© Springer International Publishing AG 2017
P. Giard et al., *High-Speed Decoders for Polar Codes*,
DOI 10.1007/978-3-319-59782-9_5

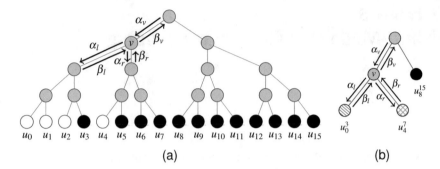

Fig. 5.1 Decoder trees for SC (**a**) and Fast-SSC (**b**) decoding of a (16, 12) polar code

mance and implementation results are provided in Sect. 5.5. Comparison against the fastest state-of-the-art polar decoder implementations in the literature is carried out in Sect. 5.5 as well. Finally, a conclusion is drawn in Sect. 5.6.

5.2   Polar Code Example and its Decoder Tree Representations

Figure 5.1a illustrates the decoder tree for a (16, 12) polar code, where black and white nodes are information and frozen bits, respectively. The Left-Hand-Side (LHS) and Right-Hand-Side (RHS) subtrees rooted in the top node are polar codes of length $N/2$. In the remainder of this chapter, we designate the polar code, of length N, decoded by traversing the whole decoder tree as the *master code* and the various codes of lengths smaller than N as *constituent codes*.

5.3   Unrolled Architectures

In an unrolled decoder, each and every operation required is instantiated so that data can flow through the decoder with minimal control. Unrolled architectures for polar decoders are described in depth in Chap. 4.

Figure 5.2 shows a fully-unrolled partially-pipelined decoder [21] with an initiation interval $\mathscr{I} = 2$ for the (16, 12) polar code of Fig. 5.1b—the initiation interval can be seen as the minimum number of clock cycles between two codeword estimates. Some control and routing logic was added to make it multi-mode as proposed in this chapter, details are provided in the next section. The α and β blocks illustrated in light blue are registers storing LLRs or bit estimates, respectively. White blocks are Fast-SSC functions as detailed in Sect. 1.6, with the exception of the "&" blocks that are concatenation operators.

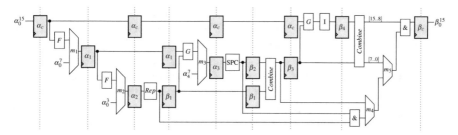

Fig. 5.2 Unrolled partially-pipelined decoder for a $(16, 12)$ polar code with initiation interval $\mathscr{I} = 2$. Clock, flip-flop enable and multiplexer select signals are omitted for clarity

5.4 Multi-Mode Unrolled Decoders

It can be noted that an unrolled decoder for a polar code of length N is composed of unrolled decoders for two polar codes of length $N/2$, which are each composed of unrolled decoders for two polar codes of length $N/4$, and so on. Thus, by adding some control and routing logic, it is possible to directly feed and read data from the unrolled decoders for constituent codes of length smaller than N. The end result is a multi-mode decoder supporting frames of various lengths and code rates.

5.4.1 Hardware Modifications to the Unrolled Decoders

Consider the decoder tree shown in Fig. 5.1b along with its unrolled implementation as illustrated in Fig. 5.2. In Fig. 5.1b, the constituent code taking root in v is an $(8, 4)$ polar code. Its corresponding decoder can be directly employed by placing the eight channel LLRs into α_0^7 and by selecting the bottom input of the multiplexer m_1 illustrated in Fig. 5.2. Its estimated codeword is retrieved from reading the output of the *Combine* block feeding the β_4 register i.e. by selecting the top and bottom inputs from m_4 and m_5, respectively, and by reading the eight least-significant bits from β_0^{15}. Similarly, still in Fig. 5.2, the decoders for the repetition and SPC constituent codes can be fed via the m_2 and m_3 multiplexers and their output eventually recovered from the output of the *Rep* and SPC blocks, respectively.

Although not illustrated in Fig. 5.2, the unrolled decoders proposed in the previous chapter feature a minimal controller. As described at the end of Sect. 4.3.3, its main task is twofold. First, it generates a done signal to indicate that a new estimated codeword is available at the output. Second, in the case of a partially-pipelined decoder i.e. with an initiation interval $\mathscr{I}$ greater than 1 like in Fig. 5.2, it asserts the various flip-flop enable signals at the correct time. Both are accomplished using a counter, albeit independently.

While not mandatory, the functionality of these counters is altered to better accommodate the use of multiple polar codes. Two LUTs are added. One LUT stores the decoding latency, in CCs, of each code. It serves as a stopping criteria to generate the done signal. The other LUT stores the clock cycle "value" i_{start} at which the enable-signal generator circuit should start. Each non-master code may start at a value $(i_{start} \mod \mathscr{I}) \neq 0$. In such cases, using the unaltered controller would result in the waste of $(i_{start} \mod \mathscr{I})$ CCs. It can be significant for short codes, especially with large values of $\mathscr{I}$. For example, without these changes, for the implementation with a master code of length 1024 and $\mathscr{I} = 20$ presented in Sect. 5.5 below, the latency for the (128, 96) polar code would increase by 20% as $(i_{start} \mod \mathscr{I}) = 17$ and the decoding latency is of 82 CCs.

Lastly, the modified controller also generates the multiplexer select signals, allowing proper data routing, based on the selected mode.

5.4.2 On the Construction of the Master Code

Conventional approaches construct polar codes for a given channel type and condition. In this work, many of the constituent codes contained within a master code are not only used internally to detect and correct errors, they are used separately as well. Therefore, we propose to assemble a master code using two optimized constituent codes in order to increase the number of optimized polar codes available. Doing so, the number of information bits, or the code rate, of the second largest supported codes can be selected. In the following, a master code of length 2048 is constructed by concatenating two constituent codes of length 1024. The LHS and RHS constituent codes are chosen to have a rate of $1/2$ and of $5/6$, respectively. As a result, the assembled master code has rate $2/3$. The location of the frozen bits in the master code is dictated by its constituent codes. Note that the constituent code with the lowest rate is put on the left—and the one with the highest rate on the right—to minimize the coding loss associated with a non-optimized polar code.

Figure 5.3 shows both the FER (left) and the BER (right) of two different (2048, 1365) polar codes. The black-solid curve is the performance of a polar code optimized using the method described in [61] for $E_b/N_0 = 4$ dB. The dashed-red curve is for the (2048, 1365) constructed by assembling (concatenating) a (1024, 512) polar code and a (1024, 853) polar code. Both polar codes of length 1024 were also optimized using the method of [61] for E_b/N_0 values of 2.5 and 5 dB, respectively.

From the figure, it can be seen that constructing an optimized polar code of length 2048 with rate $2/3$ results in a coding gain of approximately 0.17 dB at a FER of 10^{-3}—an FER appropriate for certain applications—over one assembled from two shorter polar codes of length 1024. The gap is increasing with the signal-to-noise ratio, reaching 0.24 dB at a FER of 10^{-4}. Looking at the BER curves, it can be observed that the gap is much narrower. Compared to that of the assembled master code, the optimized polar code shows a coding gain of 0.07 dB at a BER of 10^{-5}.

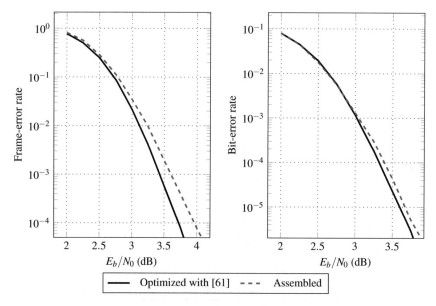

Fig. 5.3 Error-correction performance of two $(2048, 1365)$ polar codes with different constructions

5.4.3 About Constituent Codes: Frozen Bit Locations, Rate and Practicality

The location of the frozen bits in non-optimized constituent codes is dictated by their parent code. In other words, if the master code of length N has been assembled from two optimized (constituent) polar codes of length $N/2$ as suggested in the previous section, the shorter optimized codes of length $N/2$ determine the location of the frozen bits in their respective constituent codes of length $< N/2$. Otherwise, the master code dictates the frozen bit locations for all constituent codes.

Assuming that the decoding algorithm takes advantage of the a priori knowledge of these locations, the code rate and frozen bit locations of constituent codes cannot be changed at execution time. However, there are many constituent codes to choose from and code shortening can be used [38] to create more, e.g. in order to obtain a specific number of information bits or code rate.

Because of the polarization phenomenon, given any two sibling constituent codes, the code rate of the LHS one is always lower than that of the RHS one for a properly constructed polar code [58]. That property plays to our advantage as, in many wireless applications, it is desirable to offer a variety of codes of both high and low rates.

It should be noted that not all constituent codes within a master code are of practical use e.g. codes of very high rate offer negligible coding gain over an uncoded communication. For example, among the four constituent codes of length

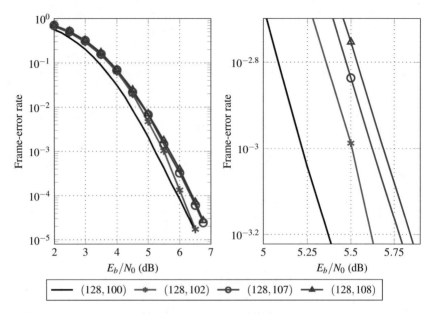

Fig. 5.4 Error-correction performance of the four constituent codes of length 128 with a rate of approximately $5/6$ contained in the proposed (2048, 1365) master code

4 included in the (16, 12) polar code illustrated in Fig. 5.1a, two of them are rate-1 constituent codes. Using them would be equivalent to uncoded communication. Moreover, among constituent codes of the same length, many codes may have a similar number of information bits with little to no error-correction performance difference in the region of interest.

Figure 5.4 shows the FER of all four constituent codes of length 128 with a rate of approximately $5/6$ that are contained within the proposed (2048, 1365) master code. It can be seen that, even at such a short length, at a FER of 10^{-3} the gap between both extremes is under 0.5 dB. Among those constituent codes, only the (128, 108) was selected for the implementation presented in Sect. 5.5. It is beneficial to limit the number of codes supported in a practical implementation of a multi-mode decoder in order to minimize routing circuitry.

5.4.4 Latency and Throughput Considerations

If a decoding algorithm taking advantage of the a priori knowledge of the frozen bit locations is used in the unrolled decoder, such as Fast-SSC [55], the latency will vary even among constituent codes of the same length. However, the coded throughput will not. The coded throughput of an unrolled decoder for a polar code of length N will be twice that of a constituent code of $N/2$, which in turn, is double that of a constituent code of length $N/4$, and so on.

In an unrolled decoder, the coded and information throughput are defined by the code length N, the clock frequency in Hz f, the initiation interval $\mathscr{I}$ and the code rate R [21]. They can be represented as

$$\mathscr{T}_C = \frac{N \cdot f}{\mathscr{I}} \quad \text{and} \quad \mathscr{T}_I = \frac{R \cdot N \cdot f}{\mathscr{I}}, \tag{5.1}$$

respectively.

In wireless communication standards where multiple code lengths and rates are supported, the peak information throughput is typically achieved with the longest code that has both the greatest latency and highest code rate. It is not mandatory to reproduce this with our proposed method, but it can be done if considered desirable. It is the example that we provide in the implementation section of this chapter.

Another possible scenario would be to use a low-rate master code, e.g. $R = 1/3$, that is more powerful in terms of error-correction performance. The resulting multi-mode decoder would reach its peak information throughput with the longest constituent code of length $N/2$ that has the highest code rate, a code with a significantly lower decoding latency than that of the master code.

5.5   Implementation Results

In this section, we present results for two implementations of our proposed multi-mode unrolled decoder with the objective of building decoders with a throughput in the vicinity of 20 Gbps. These examples are built around $(1024, 853)$ and $(2048, 1365)$ master codes. In the following, the former is referred to as the decoder supporting a maximum code length $N_{\max}$ of 1024 and the latter as the decoder with $N_{\max} = 2048$. A total of ten polar codes were selected for the decoder supporting codes of lengths up to 2048. The other decoder with $N_{\max} = 1024$ has eight modes corresponding to a subset of the ten polar codes supported by the bigger decoder. The master codes used in this section are the same as those used in Sect. 5.4.2.

For the decoder with $N_{\max} = 1024$, the Repetition and SPC nodes were constrained to a maximum size N_v of 8 and 4, respectively, as proposed in the previous chapter. For the decoder with $N_{\max} = 2048$, we found it more beneficial to lower the execution frequency and increase the maximum sizes of the Repetition and SPC nodes to 16 and 8, respectively. Additionally, the decoder with $N_{\max} = 2048$ also uses RepSPC [55] nodes to reduce latency.

ASIC synthesis results are for the 65 nm CMOS GP technology from TSMC and are obtained with Cadence RTL Compiler. Power consumption estimations are also obtained from Cadence RTL Compiler, switching activity is derived from simulation vectors. Four and five bits of quantization are used for the channel and internal LLRs, respectively. Only registers were used for memory due to the lack of access to an SRAM compiler.

We start by showing the error-correction performance of the various polar codes supported by the implementations. We later present the latency and throughput results for each of these polar codes. This section ends with synthesis results along with power consumption estimations and a comparison against the state-of-the-art polar decoder implementations.

5.5.1   Error-Correction Performance

Figure 5.5 shows the FER performance of ten different polar codes. The decoder with $N_{max} = 2048$ supports all ten illustrated polar codes whereas the decoder with $N_{max} = 1024$ supports all polar codes but the two shown as dotted curves. All simulations are generated using random codewords modulated with BPSK and transmitted over an AWGN channel.

It can be seen from the figure that the error-correction performance of the supported polar codes varies greatly. As expected, for codes of the same lengths, the codes with the lowest code rates performs significantly better than their higher

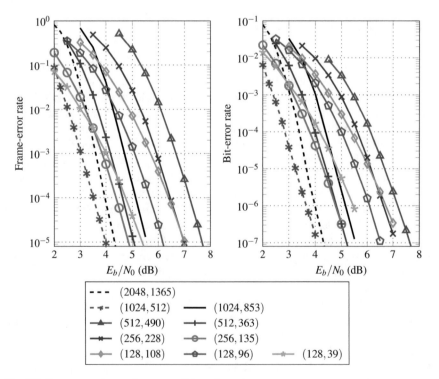

Fig. 5.5 Error-correction performance of the polar codes

rate counterpart. For example, at a FER of 10^{-4}, the performance of the $(512, 363)$ polar code is almost 3 dB better than that of the $(512, 490)$ code.

While the error-correction performance plays a role in the selection of a code, the latency and throughput are also important considerations. As it will be shown in the following section, the ten selected polar codes perform much differently in that regard as well.

5.5.2 Latency and Throughput

Table 5.1 shows the latency and information throughput for both decoders with $N_{max} \in \{1024, 2048\}$. To reduce the area and latency while retaining the same throughput, the initiation interval $\mathscr{I}$ can be increased along with the clock frequency (5.1) [21].

If both decoders have initiation intervals of 20—as used in the section below— Table 5.1 assumes clock frequencies of 500 and 250 MHz for the decoders with $N_{max} = 1024$ and $N_{max} = 2048$, respectively. While their master codes differ, both decoders feature a peak information throughput in the vicinity of 20 Gbps. For the decoder with the smallest N_{max}, the seven other polar codes have an information throughput in the multi-gigabit per second range with the exception of the shortest and lowest-rate constituent code. That $(128, 39)$ constituent code still has an information throughput close to 1 Gbps. The decoder with $N_{max} = 2048$ offers multi-gigabit throughput for most of the supported polar codes. The minimum information throughput is also with the $(128, 39)$ polar code at approximately 500 Mbps.

In terms of latency, the decoder with $N_{max} = 1024$ requires 646 ns to decode its longest supported code. The latency for all the other codes supported by that decoder is under 500 ns. Even with its additional dedicated node and relaxed maximum size constraint on the Repetition and SPC nodes, the decoder with $N_{max} = 2048$ has greater latency overall because of its lower clock frequency. For example, its latency is of 2.01 μs, 944 ns and 1.06 μs for the $(2048, 1365)$, $(1024, 853)$ and $(1024, 512)$ polar codes, respectively.

Using the same nodes and constraints as for $N_{max} = 1024$, the $N_{max} = 2048$ decoder would allow for greater clock frequencies. While 689 CCs would be required to decode the longest polar code instead of 503, a clock of 500 MHz would be achievable, effectively reducing the latency from 2.01 to 1.38 μs and doubling the throughput. However, this reduction comes at the cost of much greater area and an estimated power consumption close to 1 W.

Table 5.1 Information throughput and latency for the multi-mode unrolled polar decoders based on the (2048, 1365) and (1024, 853) master codes, respectively with a N_{max} of 1024 and 2048

Code (N, k)	Rate (k/N)	Info. T/P (Gbps)		Latency (CCs)		Latency (ns)	
		$N_{max} = 1024$	$N_{max} = 2048$	$N_{max} = 1024$	$N_{max} = 2048$	$N_{max} = 1024$	$N_{max} = 2048$
(2048, 1365)	2/3	–	17.1	–	503	–	2012
(1024, 853)	5/6	21.3	10.7	323	236	646	944
(1024, 512)	1/2	–	6.4	–	265	–	1060
(512, 490)	19/20	12.3	6.2	95	75	190	300
(512, 363)	7/10	9.1	4.5	226	159	452	636
(256, 228)	9/10	5.7	2.6	86	61	172	244
(256, 135)	1/2	3.4	1.7	138	96	276	384
(128, 108)	5/6	2.7	1.4	54	40	108	160
(128, 96)	3/4	2.4	1.2	82	52	164	208
(128, 39)	1/3	0.98	0.49	54	42	108	168

5.5.3 Synthesis Results and Comparison with the State of the Art

Table 5.2 shows the synthesis results along with power consumption estimations for the two implementations of the proposed multi-mode unrolled decoder. The work in the first two columns is for the decoder with $N_{max} = 1024$, based on the $(1024, 853)$ master code. It was synthesized for clock frequencies of 500 and 650 MHz, respectively, with initiation intervals $\mathscr{I}$ of 20 and 26. Our work shown in the third and fourth columns is for the decoders with $N_{max} = 2048$, built from the assembled $(2048, 1365)$ polar code. These decoders have an initiation interval $\mathscr{I}$ of 20 or 28, with lower clock frequencies of 250 and 350 MHz, respectively. For comparison with other works, the same table also includes results for a dedicated partially-pipelined decoder for a $(1024, 512)$ polar code as presented in Chap. 4.

The four fastest polar decoder implementations from the literature are also included for comparison along with normalized area results. For consistency, only the largest polar code supported by each of our proposed multi-mode unrolled decoders is used and the coded throughput, as opposed to the information one, is compared to match what was done in most of the other works.

From Table 5.2, it can be seen that the area for the proposed decoders with $N_{max} = 1024$ are similar to that of the BP decoder of [49] as well as the normalized area for the unrolled SC decoder from [14]. However, their area is from 2.1 to 2.5 times greater than that of [17]. Comparing the multi-mode decoders, the area for the decoder with $N_{max} = 2048$ is over twice that of the ones with $N_{max} = 1024$, however the master code for the former has twice the length of the latter and supports two more modes.

All proposed decoders have a coded throughput that is an order of magnitude greater than the other works. Latency is approximately the same as that of the BP decoder for $N_{max} = 1024$, and from three to four times greater with $N_{max} = 2048$. Comparing against the SC decoder of [14], the latency is 1.7 or 3.7 times greater for decoders with an N_{max} of 1024 and 2048, respectively. It should be noted that the decoder of [14] support codes of any rate, where the proposed multi-mode decoders support a limited number of code rates.

The latency of the proposed decoders is higher than the programmable Fast-SSC decoder of [17]. This is due to greater limitations on the specialized repetition and SPC decoders. The decoder in [17] limits repetition decoders to a maximum length of 32, compared to 8 or 16 in this work, and does not place limits on the SPC decoders.

Finally, among the decoders with $N_{max} = 1024$ implemented in 65 nm with a power supply of 1 V and operating at 25 °C, our proposed implementation offers the greatest area and energy efficiency. The proposed multi-mode decoder exhibits 3.3 and 5.6 times better area efficiency than the decoders of [17] and [49], respectively. The energy efficiency is estimated to be 2.7 and 4.8 times higher compared to that of the same two decoders from the literature.

Table 5.2 Comparison with state-of-the-art polar decoders

	Multi-mode				Dedicated	[17]	[49]ª	[14]	[72]
Algorithm	Fast-SSC				Fast-SSC	Fast-SSC	BP	SC	2-bit SC
Technology	65 nm				65 nm	65 nm	65 nm	90 nm	45 nm
N_{max}	1024		2048		1024	1024	1024	1024	1024
Code	(1024, 853)		(2048, 1365)		(1024, 512)	(1024, 512)	(1024, 512)	(1024, k)	(1024, 512)
Init. interval ($\mathscr{I}$)	20	26	20	28	20	–	–	–	–
Supply (V)	0.72	1.0	0.72	1.0	1.0	1.0	1.0	1.3	N/A
Oper. temp. (°C)	125	25	125	25	25	25	≈25	N/A	N/A
Area (mm²)	1.71	1.44	4.29	3.58	1.68	0.69	1.48	3.21	N/A
Area @65 nm (mm²)	1.71	1.44	4.29	3.58	1.68	0.69	1.48	1.68	0.4
Frequency (MHz)	500	650	250	350	500	600	300	2.5	750
Latency (µs)	0.65	0.50	2.01	1.44	0.73	0.27	0.50	0.39	1.02
Coded T/P (Gbps)	25.6	25.6	25.6	25.6	25.6	3.7	4.7 @ 4 dB	2.56	1.0
Sust. coded T/P (Gbps)	25.6	25.6	25.6	25.6	25.6	3.7	2.0	2.56	1.0
Area Eff. (Gbps/mm²)	15.42	17.75	5.97	7.16	15.27	5.40	3.18 @ 4 dB	0.80	N/A
Power (mW)	226	546	379	740	386	215	478	191	N/A
Energy (pJ/bit)	8.8	21.3	14.8	28.9	15.1	57.7	102.1	74.5	N/A

ª Measurement results

Recently, a List-based multi-mode decoder was proposed in [71], where the definition of the word "multi-mode" differs greatly with our work: in our work, it is used to indicate that the decoder is capable of decoding codes with varying length and rate. Whereas in [71], a "mode" indicates the level of parallelism in the decoder. The decoder of [71] is capable of decoding four paths in parallel by implementing four processing units. It can be configured to either do SC-based decoding of four frames or List-based decoding. For the latter, two list sizes L are supported. If $L = 2$, two frames are decoded in parallel otherwise if $L = 4$, only one frame is decoded at a time.

5.6 Conclusion

In this chapter we presented a new method to transform an unrolled architecture into a multi-mode decoder supporting various polar code lengths and rates. We showed that a master code can be assembled from two optimized polar codes of smaller length, with desired code rates, without sacrificing too much coding gain. We provided results for two decoders, one built for a $(1024, 853)$ master code and the other for a longer $(2048, 1365)$ polar code. Both decoders support from seven to nine other practical codes. On 65 nm ASIC, they were shown to have a peak throughput greater than 25 Gbps. One has a worst-case latency of $2\,\mu s$ at 250 MHz and an energy efficiency of 14.8 pJ/bit. The other has a worst-case latency of 646 ns at 500 MHz and an energy efficiency of 8.8 pJ/bit. Both implementation examples show that, with their great throughput and support for codes of various lengths and rates, multi-mode unrolled polar decoders are promising candidates for future wireless communication standards.

Chapter 6
Conclusion and Future Work

Abstract Conclusions about this book are drawn in this chapter and a list of suggested future research topics is presented.

Error-correcting codes play a crucial role in reliable and robust communication and storage systems. The dream of researchers would be to achieve the channel capacity at low implementation complexity without compromising the latency and throughput requirements of living a modern connected life. Polar codes are the latest class of modern error-correcting codes and they show high potential. The early decoder implementations greatly suffered from high latency and low throughput. The state-of-the-art low-complexity algorithm improved the situation but contained improvements targeted at high-rate codes and the throughput was still an order of magnitude lower than the expected requirements for future wireless communication standards. Initially there was a lack of software implementations suitable for high-performance SDR applications and then the work that appeared suffered from a high latency and memory footprint. This book presented solutions to address those issues.

The optimization presented in the original Fast-SSC algorithm [55], the fastest low-complexity decoding algorithm, targeted high-rate codes. In Chap. 2, we showed how to improve the Fast-SSC algorithm by adding dedicated decoders for three new types of constituent codes frequently appearing in low-rate codes. We also introduced a human-guided polar code construction alteration method to significantly reduce the latency and increase the throughput of a Fast-SSC decoder at the cost of a small error-correction performance loss. The resulting decoders for codes of rate $1/2$ and $1/3$ presented in this work achieved an information throughput greater 1.2 Gbps at an operating frequency of 400 MHz, while retaining the low complexity of the original Fast-SSC implementation.

With their low-complexity encoding and decoding algorithms, polar codes are attractive for applications where computational resources are limited and a custom hardware solution too costly. Chapter 3 presented low-latency software polar decoders exploiting the capabilities offered in modern processors. By adapting the algorithms at various levels, the software decoders presented in this work had an order of magnitude lower latency and memory footprint compared to the state-of-the-art decoders, while maintaining a comparable throughput. In addition,

P. Giard et al., *High-Speed Decoders for Polar Codes*,
DOI 10.1007/978-3-319-59782-9_6

we presented strategies for implementing polar decoders on graphical processing units and showed that hundreds of frames could be simultaneously decoded while sustaining a throughput greater than 1 Gbps.

Chapter 4 introduced a family of hardware architectures using a reduced-complexity successive-cancellation decoding algorithm that employs unrolling. It demonstrated that polar decoders can achieve extremely high throughput values and retain moderate complexity. The resulting fully-unrolled architectures were shown to be capable of achieving a throughput that is two to three orders of magnitude greater than current state-of-the-art polar decoders, while retaining a good energy efficiency.

Many communication standards mandate the error-correction system to support various code lengths and rate in order to adapt to varying channel conditions or latency requirements. Multi-mode unrolled hardware architectures and implementations were proposed in Chap. 5. This novel method lead to a length- and rate-flexible decoder while retaining the very high speed typical to unrolled decoders. Results were presented for two versions of a multi-mode decoder supporting eight and ten different polar codes, respectively. Both showed that, with their throughput greater than 25 Gbps, latency below 2 μs and support for codes of various lengths and rates, multi-mode unrolled polar decoders are promising candidates for future wireless communication standards.

6.1   Future Work

The research presented in this book showed that polar codes are a new class of modern error-correcting codes that already show great potential for use in some practical applications. For example, encoding and decoding of polar codes on modern processors for use in SDR applications already makes sense. However, the software implementations we presented were not suitable for micro-controller processors omnipresent in Internet of Things (IoT) devices. Also, as the error-correction performance of moderate-length polar codes—when an SC-based decoding algorithm is used—is less than that of LDPC codes, polar codes will not take over the world of error correction just yet. List-based decoding can close that gap [62] but its hardware implementations suffer from low throughput [9, 36] even if there is hope for improvements [24, 56]. Here is a list of suggested future research topics that would help broaden the scope of interesting applications for polar codes.

6.1.1   Software Encoding and Decoding on APU Processors

The GPU implementation results presented in Chap. 3 showed that hundreds of frames could be simultaneously decoded at a sustained throughput greater than 1 Gbps. That throughput was shown to be I/O bound to the capabilities of the PCIe bus. Even with a GPU and motherboard supporting a faster, more recent, iteration of the PCIe, the memory copy latency for moving data from the host memory to the card memory will remain a significant barrier to a better throughput. There exists processors coupled with a GPGPU on the same die sharing the same memory. AMD's APU are among those. It would be interesting to investigate the use of APUs to conduct software decoding of polar codes.

6.1.2   Software Encoding and Decoding on Micro-Controllers

A great share of the optimization strategies presented in Chap. 3 cannot be applied to micro-controllers, processors that do not have SIMD instructions. Most IoT devices of today either use micro-controllers or a SoC that features one because of their relative low cost. While these devices are not powerful enough to implement a practical LDPC or turbo decoder, an error-correction solution based on polar codes instead of a classic codes such as BCH or RS codes is certainly an interesting avenue to explore.

6.1.3   Multi-Mode Unrolled List Decoders

The focus is shifting away from SC-based decoding in favor of List-based decoding as CRC-aided List decoding of polar codes can outperform the error-correction performance of LDPC codes. However, the current state of the art in hardware List decoders suffers from a low throughput and high latency. The multi-mode unrolled hardware architectures in Chap. 5 can be applied to List-based decoding. In order to keep complexity practical, it would be interesting to adapt and implement such architectures for small list sizes.

References

1. PCI express base specification revision 2.0. PCI-SIG (2006)
2. IEEE standard for floating-point arithmetic. IEEE Std 754-2008 pp. 1–70 (2008). doi:10.1109/IEEESTD.2008.4610935
3. IEEE standard for information technology–telecommunications and information exchange between systems local and metropolitan area networks–specific requirements part 11: Wireless LAN medium access control (MAC) and physical layer (PHY) specifications. IEEE Std 802.11-2012 (Revision of IEEE Std 802.11-2007) pp. 1–2793 (2012). doi:10.1109/IEEESTD.2012.6178212
4. Alamdar-Yazdi, A., Kschischang, F.R.: A simplified successive-cancellation decoder for polar codes. IEEE Commun. Lett. **15**(12), 1378–1380 (2011). doi:10.1109/LCOMM.2011. 101811.111480
5. Altera: Meeting the performance and power imperative of the zettabyte era with generation 10. White Paper (2013)
6. Arıkan, E.: Channel polarization: A method for constructing capacity-achieving codes. In: IEEE Int. Symp. on Inf. Theory (ISIT), pp. 1173–1177 (2008). doi:10.1109/ISIT.2008.4595172
7. Arıkan, E.: Channel polarization: A method for constructing capacity-achieving codes for symmetric binary-input memoryless channels. IEEE Trans. Inf. Theory **55**(7), 3051–3073 (2009). doi:10.1109/TIT.2009.2021379
8. Arıkan, E.: Systematic polar coding. IEEE Commun. Lett. **15**(8), 860–862 (2011). doi:10.1109/LCOMM.2011.061611.110862
9. Balatsoukas-Stimming, A., Bastani Parizi, M., Burg, A.: LLR-based successive cancellation list decoding of polar codes. IEEE Trans. Signal Process. **63**(19), 5165–5179 (2015). doi:10.1109/TSP.2015.2439211
10. Balatsoukas-Stimming, A., Karakonstantis, G., Burg, A.: Enabling complexity-performance trade-offs for successive cancellation decoding of polar codes. In: IEEE Int. Symp. on Inf. Theory (ISIT), pp. 2977–2981 (2014). doi:10.1109/ISIT.2014.6875380
11. Bang, S., Ahn, C., Jin, Y., Choi, S., Glossner, J., Ahn, S.: Implementation of LTE system on an SDR platform using CUDA and UHD. Analog Integr. Circuits and Signal Process. **78**(3), 599–610 (2014). doi:10.1007/s10470-013-0229-1
12. Berrou, C., Glavieux, A., Thitimajshima, P.: Near Shannon limit error-correcting coding and decoding: Turbo-codes. In: IEEE Int. Conf. Commun. (ICC), vol. 2, pp. 1064–1070 (1993). doi:10.1109/ICC.1993.397441
13. Demel, J., Koslowski, S., Jondral, F.: A LTE receiver framework using GNU Radio. J. Signal Process. Syst. **78**(3), 313–320 (2015). doi:10.1007/s11265-014-0959-z

© Springer International Publishing AG 2017

P. Giard et al., *High-Speed Decoders for Polar Codes*,

DOI 10.1007/978-3-319-59782-9

14. Dizdar, O., Arıkan, E.: A high-throughput energy-efficient implementation of successive cancellation decoder for polar codes using combinational logic. IEEE Trans. Circuits Syst. I **63**(3), 436–447 (2016). doi:10.1109/TCSI.2016.2525020

15. Feng, W.C., Xiao, S.: To GPU synchronize or not GPU synchronize? In: IEEE Int. Symp. on Circuits and Syst. (ISCAS), pp. 3801–3804 (2010). doi:10.1109/ISCAS.2010.5537722

16. Gallager, R.: Low-density parity-check codes. IRE Trans. Inf. Theory **8**(1), 21–28 (1962). doi:10.1109/TIT.1962.1057683

17. Giard, P., Balatsoukas-Stimming, A., Sarkis, G., Thibeault, C., Gross, W.J.: Fast low-complexity decoders for low-rate polar codes. Springer J. Signal Process. Syst. (2016). doi:10.1007/s11265-016-1173-y

18. Giard, P., Sarkis, G., Thibeault, C., Gross, W.J.: Fast software polar decoders. In: IEEE Int. Conf. on Acoustics, Speech, and Signal Process. (ICASSP), pp. 7555–7559 (2014). doi:10.1109/ICASSP.2014.6855069

19. Giard, P., Sarkis, G., Thibeault, C., Gross, W.J.: 237 Gbit/s unrolled hardware polar decoder. IET Electron. Lett. **51**(10), 762–763 (2015). doi:10.1049/el.2014.4432

20. Giard, P., Sarkis, G., Thibeault, C., Gross, W.J.: A 638 Mbps low-complexity rate 1/2 polar decoder on FPGAs. In: IEEE Int. Workshop on Signal Process. Syst. (SiPS), pp. 1–6 (2015). doi:10.1109/SiPS.2015.7345007

21. Giard, P., Sarkis, G., Thibeault, C., Gross, W.J.: Multi-mode unrolled hardware architectures for polar decoders. IEEE Trans. Circuits Syst. I **63**(9), 1443–1453 (2016). doi:10.1109/TCSI.2016.2586218

22. Goela, N., Korada, S.B., Gastpar, M.: On lp decoding of polar codes. In: IEEE Inf. Theory Workshop (ITW), pp. 1–5 (2010). doi:10.1109/CIG.2010.5592698

23. Han, X., Niu, K., He, Z.: Implementation of IEEE 802.11n LDPC codes based on general purpose processors. In: IEEE Int. Conf. on Commun. Technol. (ICCT), pp. 218–222 (2013). doi:10.1109/ICCT.2013.6820375

24. Hashemi, S.A., Balatsoukas-Stimming, A., Giard, P., Thibeault, C., Gross, W.J.: Partitioned successive-cancellation list decoding of polar codes. In: IEEE Int. Conf. on Acoustics, Speech, and Signal Process. (ICASSP), pp. 957–960 (2016). doi:10.1109/ICASSP.2016.7471817

25. Huang, Z., Diao, C., Chen, M.: Latency reduced method for modified successive cancellation decoding of polar codes. IET Electron. Lett. **48**(23), 1505–1506 (2012). doi:10.1049/el.2012.2795

26. Hussami, N., Urbanke, R., Korada, S.B.: Performance of polar codes for channel and source coding. In: IEEE Int. Symp. on Inf. Theory (ISIT), pp. 1488–1492 (2009). doi:10.1109/ISIT.2009.5205860

27. Jouguet, P., Kunz-Jacques, S.: High performance error correction for quantum key distribution using polar codes. Quantum Inf. & Computation **14**(3-4), 329–338 (2014)

28. Kahraman, S., Çelebi, M.E.: Code based efficient maximum-likelihood decoding of short polar codes. In: IEEE Int. Symp. on Inf. Theory (ISIT), pp. 1967–1971 (2012). doi:10.1109/ISIT.2012.6283643

29. Karjalainen, J., Nekovee, M., Benn, H., Kim, W., Park, J., Sungsoo, H.: Challenges and opportunities in mm-wave communication in 5G networks. In: Int. Conf. on Cognitive Radio Oriented Wireless Netw. and Commun. (CROWNCOM), pp. 372–376 (2014). doi:10.4108/icst.crowncom.2014.255604

30. Le Gal, B., Jego, C., Crenne, J.: A high throughput efficient approach for decoding LDPC codes onto GPU devices. IEEE Embedded Syst. Lett. **6**(2), 29–32 (2014). doi:10.1109/LES.2014.2311317

31. Le Gal, B., Leroux, C., Jego, C.: Software polar decoder on an embedded processor. In: IEEE Int. Workshop on Signal Process. Syst. (SiPS) (2014). doi:10.1109/SiPS.2014.6986083

32. Le Gal, B., Leroux, C., Jego, C.: Multi-Gb/s software decoding of polar codes. IEEE Trans. Signal Process. **63**(2), 349–359 (2015). doi:10.1109/TSP.2014.2371781

33. Leroux, C., Raymond, A., Sarkis, G., Gross, W.: A semi-parallel successive-cancellation decoder for polar codes. IEEE Trans. Signal Process. **61**(2), 289–299 (2013). doi:10.1109/TSP.2012.2223693

34. Leroux, C., Raymond, A.J., Sarkis, G., Tal, I., Vardy, A., Gross, W.J.: Hardware implementation of successive-cancellation decoders for polar codes. J. Signal Process. Syst. **69**(3), 305–315 (2012). doi:10.1007/s11265-012-0685-3

35. Leroux, C., Tal, I., Vardy, A., Gross, W.J.: Hardware architectures for successive cancellation decoding of polar codes. In: IEEE Int. Conf. on Acoust., Speech and Signal Process. (ICASSP), pp. 1665–1668 (2011). doi:10.1109/ICASSP.2011.5946819

36. Li, B., Shen, H., Tse, D.: An adaptive successive cancellation list decoder for polar codes with cyclic redundancy check. IEEE Commun. Lett. **16**(12), 2044–2047 (2012). doi:10.1109/LCOMM.2012.111612.121898

37. Li, B., Shen, H., Tse, D., Tong, W.: Low-latency polar codes via hybrid decoding. In: Int. Symp. on Turbo Codes and Iterative Inf. Process. (ISTC), pp. 223–227 (2014). doi:10.1109/ISTC.2014.6955118

38. Li, Y., Alhussien, H., Haratsch, E., Jiang, A.: A study of polar codes for MLC NAND flash memories. In: Int. Conf. on Comput., Netw. and Commun. (ICNC), pp. 608–612 (2015). doi:10.1109/ICCNC.2015.7069414

39. MacKay, D.J.C., Neal, R.M.: Near shannon limit performance of low density parity check codes. IET Electron. Lett. **33**(6), 457–458 (1997). doi:10.1049/el:19970362

40. MCC Support: Final Report of 3GPP TSG RAN WG1 #87 v1.0.0 (2017)

41. Miloslavskaya, V.: Shortened polar codes. IEEE Trans. Inf. Theory **61**(9), 4852–4865 (2015). doi:10.1109/TIT.2015.2453312

42. Mishra, A., Raymond, A., Amaru, L., Sarkis, G., Leroux, C., Meinerzhagen, P., Burg, A., Gross, W.: A successive cancellation decoder ASIC for a 1024-bit polar code in 180nm CMOS. In: IEEE Asian Solid State Circuits Conf. (A-SSCC), pp. 205–208 (2012). doi:10.1109/IPEC.2012.6522661

43. Monserrat, J.F., Mange, G., Braun, V., Tullberg, H., Zimmermann, G., Bulakci, Ö.: METIS research advances towards the 5G mobile and wireless system definition. EURASIP J. Wireless Commun. Netw. **2015**(1), 1–16 (2015). doi:10.1186/s13638-015-0302-9

44. Mori, R., Tanaka, T.: Performance and construction of polar codes on symmetric binary-input memoryless channels. In: IEEE Int. Symp. on Inf. Theory (ISIT), pp. 1496–1500 (2009). doi:10.1109/ISIT.2009.5205857

45. NVIDIA: Kepler GK110 - the fastest, most efficient HPC architecture ever built. NVIDIA's Next Generation CUDA Computer Architecture: Kepler GK110 (2012)

46. NVIDIA: NVIDIA management library (NVML). NVML API Reference Guide (2014)

47. NVIDIA: Performance guidelines. CUDA C Programming Guide (2014)

48. Pamuk, A., Arıkan, E.: A two phase successive cancellation decoder architecture for polar codes. In: IEEE Int. Symp. on Inf. Theory (ISIT), pp. 1–5 (2013). doi:10.1109/ISIT.2013.6620368

49. Park, Y.S., Tao, Y., Sun, S., Zhang, Z.: A 4.68Gb/s belief propagation polar decoder with bit-splitting register file. In: Symp. on VLSI Circ. Dig. of Tech. Papers, pp. 1–2 (2014). doi:10.1109/VLSIC.2014.6858413

50. Raymond, A.J.: Design and hardware implementation of decoder architectures for polar codes. Master's thesis, McGill University (2014)

51. Raymond, A.J., Gross, W.J.: Scalable successive-cancellation hardware decoder for polar codes. In: IEEE Glob. Conf. on Signal and Inf. Process. (GlobalSIP), pp. 1282–1285 (2013). doi:10.1109/GlobalSIP.2013.6737143

52. Roh, W.: 5G mobile communications for 2020 and beyond - vision and key enabling technologies. IEEE Wireless Commun. and Netw. Conf. (WCNC) (2014)

53. Sarkis, G.: Efficient encoders and decoders for polar codes: Algorithms and implementations. Ph.D. thesis, McGill University (2016)

54. Sarkis, G., Giard, P., Thibeault, C., Gross, W.J.: Autogenerating software polar decoders. In: IEEE Global Conf. on Signal and Inf. Process. (GlobalSIP), pp. 6–10 (2014). doi:10.1109/GlobalSIP.2014.7032067

55. Sarkis, G., Giard, P., Vardy, A., Thibeault, C., Gross, W.J.: Fast polar decoders: Algorithm and implementation. IEEE J. Sel. Areas Commun. **32**(5), 946–957 (2014). doi:10.1109/JSAC.2014.140514

56. Sarkis, G., Giard, P., Vardy, A., Thibeault, C., Gross, W.J.: Fast list decoders for polar codes. IEEE J. Sel. Areas Commun. - Special Issue on Recent Advances In Capacity Approaching Codes **34**(2), 318–328 (2016). doi:10.1109/JSAC.2015.2504299

57. Sarkis, G., Gross, W.J.: Increasing the throughput of polar decoders. IEEE Commun. Lett. **17**(4), 725–728 (2013). doi:10.1109/LCOMM.2013.021213.121633

58. Sarkis, G., Tal, I., Giard, P., Vardy, A., Thibeault, C., Gross, W.J.: Flexible and low-complexity encoding and decoding of systematic polar codes. IEEE Trans. Commun. **64**(7), 2732–2745 (2016). doi:10.1109/TCOMM.2016.2574996

59. Schläfer, P., Wehn, N., Alles, M., Lehnigk-Emden, T.: A new dimension of parallelism in ultra high throughput LDPC decoding. In: IEEE Workshop on Signal Process. Syst. (SiPS), pp. 153–158 (2013). doi:10.1109/SiPS.2013.6674497

60. Shannon, C.: A mathematical theory of communication. Bell Syst. Tech. J. **27**(3), 379–423 (1948). doi:10.1002/j.1538-7305.1948.tb01338.x

61. Tal, I., Vardy, A.: How to construct polar codes. IEEE Trans. Inf. Theory **59**(10), 6562–6582 (2013). doi:10.1109/TIT.2013.2272694

62. Tal, I., Vardy, A.: List decoding of polar codes. IEEE Trans. Inf. Theory **61**(5), 2213–2226 (2015). doi:10.1109/TIT.2015.2410251

63. Tan, K., Liu, H., Zhang, J., Zhang, Y., Fang, J., Voelker, G.M.: Sora: High-performance software radio using general-purpose multi-core processors. Commun. ACM **54**(1), 99–107 (2011). doi:10.1145/1866739.1866760

64. Treibig, J., Hager, G., Wellein, G.: LIKWID: A lightweight performance-oriented tool suite for x86 multicore environments. In: Int. Conf. on Parallel Process. Workshops (ICPPW), pp. 207–216 (2010). doi:10.1109/ICPPW.2010.38

65. Trifonov, P.: Efficient design and decoding of polar codes. IEEE Trans. Commun. **60**(11), 3221–3227 (2012). doi:10.1109/TCOMM.2012.081512.110872

66. Wang, G., Wu, M., Yin, B., Cavallaro, J.R.: High throughput low latency LDPC decoding on GPU for SDR systems. In: IEEE Glob. Conf. on Sign. and Inf. Process. (GlobalSIP), pp. 1258–1261 (2013). doi:10.1109/GlobalSIP.2013.6737137

67. Wang, R., Liu, R.: A novel puncturing scheme for polar codes. IEEE Commun. Lett. **18**(12), 2081–2084 (2014). doi:10.1109/LCOMM.2014.2364845

68. Wehn, N., Scholl, S., Schläfer, P., Lehnigk-Emden, T., Alles, M.: Challenges and limitations for very high throughput decoder architectures for soft-decoding. In: C. Chavet, P. Coussy (eds.) Advanced Hardware Design for Error Correcting Codes, pp. 7–31. Springer International Publishing (2015). doi:10.1007/978-3-319-10569-7_2

69. Xianjun, J., Canfeng, C., Jaaskelainen, P., Guzma, V., Berg, H.: A 122Mb/s turbo decoder using a mid-range GPU. In: Int. Wireless Commun. and Mobile Comput. Conf. (IWCMC), pp. 1090–1094 (2013). doi:10.1109/IWCMC.2013.6583709

70. Xilinx: UltraScale architecture and product overview. Product Specification (2014)

71. Xiong, C., Lin, J., Yan, Z.: A multimode area-efficient SCL polar decoder. IEEE Trans. VLSI Syst. **PP**(99), 1–14 (2016). doi:10.1109/TVLSI.2016.2557806

72. Yuan, B., Parhi, K.: Low-latency successive-cancellation polar decoder architectures using 2-bit decoding. IEEE Trans. Circuits Syst. I **61**(4), 1241–1254 (2014). doi:10.1109/TCSI.2013.2283779

73. Yuan, B., Parhi, K.K.: Early stopping criteria for energy-efficient low-latency belief-propagation polar code decoders. IEEE Trans. Signal Process. **62**(24), 6496–6506 (2014). doi:10.1109/TSP.2014.2366712

74. Zhang, L., Zhang, Z., Wang, X., Zhong, C., Ping, L.: Simplified successive-cancellation decoding using information set reselection for polar codes with arbitrary blocklength. IET Communications **9**(11), 1380–1387 (2015). doi:10.1049/iet-com.2014.0988

Index

© Springer International Publishing AG 2017

P. Giard et al., *High-Speed Decoders for Polar Codes*,

DOI 10.1007/978-3-319-59782-9

Printed in the United States
By Bookmasters